T0325398

Integrability using the
Sine-Gordon and Thirring Duality

An introductory course

Online at: https://doi.org/10.1088/978-0-7503-5899-6

Integrability using the
Sine-Gordon and Thirring Duality

An introductory course

Alessandro Torrielli

School of Mathematics and Physics, University of Surrey, Guildford, Surrey GU2 7XH, UK

IOP Publishing, Bristol, UK

ISBN   978-0-7503-5899-6 (ebook)
ISBN   978-0-7503-5897-2 (print)
ISBN   978-0-7503-5900-9 (myPrint)
ISBN   978-0-7503-5898-9 (mobi)

DOI   10.1088/978-0-7503-5899-6

Version: 20240501

IOP ebooks

British Library Cataloguing-in-Publication Data: A catalogue record for this book is available from the British Library.

Published by IOP Publishing, wholly owned by The Institute of Physics, London

IOP Publishing, No.2 The Distillery, Glassfields, Avon Street, Bristol, BS2 0GR, UK

US Office: IOP Publishing, Inc., 190 North Independence Mall West, Suite 601, Philadelphia, PA 19106, USA

This book is dedicated to my family, to my friends, and to my colleagues. A special dedication to all my collaborators of many years, for innumerable discussions and wonderful delightful cooperation.

Contents

Acknowledgements

I sincerely thank the organisers of the LonTi programme, and in particular Bogdan Stefański, Nadav Drukker and Yang-Hui He, for the invitation to give the lectures which have inspired the compilation of this book, for the encouragement and for the very interesting discussions at LIMS; Elli Heyes, Dmitrii Riabchenko, Chawakorn Maneerat, and the technical and video support staff for their kind help during the lectures; and the Royal Institution for hosting the lectures in their inspiring London setting. I would like to thank Suvajit Majumder for several discussions. Many thanks go to all the participants to the lectures, for their enthusiasm, their error-spotting ability and their extremely good questions, all of which have prompted several revisions of this manuscript. I wish to very much thank in addition Zoltan Bajnok, Patrick Dorey, Julius Julis, Sergey Frolov, Ctirad Klimčík, Michele Mazzoni, Ingo Runkel, Evgeny Sklyanin and Benoit Vicedo for extremely interesting and illuminating discussions. Support from the EPSRC-SFI grant EP/S020888/1 *Solving Spins and Strings* is gratefully acknowledged.

Author biography

Alessandro Torrielli

 **Alessandro Torrielli** graduated in Physics from the University of Genova (Italy) in 1999 and took his PhD in Physics from the University of Padova (Italy) in 2003. He then held postdoctoral positions at Padova, the Humboldt University of Berlin (Germany), the Massachusetts Institute of Technology (MIT, USA), the University of Utrecht (The Netherlands), and the University of York (United Kingdom). Since 2011, he has been a permanent member of staff at the University of Surrey (United Kingdom). In 2022, he was promoted to Professor of Mathematics. His research interests focus on quantum integrable systems, particularly those appearing in the context of the AdS/CFT correspondence in string theory. His research papers—with his collaborators and solo—have gathered over 3800 citations (source: Google Scholar, January 2024), and he has written and co-authored several reviews, which are well-established in his field. In 2019, he was awarded a major EPSRC-SFI research grant in cooperation with Dr Marius de Leeuw at Trinity College Dublin (Ireland).

IOP Publishing

Integrability using the Sine-Gordon and Thirring Duality
An introductory course
Alessandro Torrielli

Chapter 1

Introduction

Quantum field theories in $3 + 1$ dimensions are not normally solvable in closed analytic form. They represent an incredibly accurate description of elementary particle physics, but they remain for the most part frustratingly complex. Integrable quantum field theories are a domain of investigation where one can make exact analytic predictions. However, the price one must pay is that the dynamics must be sufficiently simplified. These theories normally reside in $1 + 1$ dimensions, and therefore play a key role as toy models: they are simple enough to allow a complete description but still exhibit a rich range of effects that are a not-too-distant counterpart of the phenomena in $3 + 1$ dimensions. Integrable field theories are essentially defined by their exact S-matrices and the associated analytic structure. Such S-matrices are calculable in closed form from first physical principles, such as unitarity, crossing symmetry and the position of the bound states. Thanks to their huge amount of symmetry, such models are an ideal playground for understanding the subtleties of nonperturbative quantisation. The theory of integrable systems also plays an extremely important role in modern advances in string theory and the so-called Anti de Sitter/conformal field theory (AdS/CFT) correspondence—see for instance [1].

Chief examples of integrable systems that exhibit a very rich content and extremely interesting dynamical features are provided by the $1 + 1$-dimensional Sine-Gordon model and the $1 + 1$-dimensional Thirring model. These two models are dual to each other, in a sense first elucidated by the pioneering work of S. Coleman [18]. Having control of exact analytic expressions is of paramount importance to be able to understand this duality at a nonperturbative level, and in general to appreciate the fine details behind the quantisation of systems with an infinite tower of conserved quantities.

One of the goals of this book is to provide a detailed description of the duality between these two integrable systems: the $1 + 1$-dimensional Sine-Gordon model and the $1 + 1$-dimensional Thirring model. While of great importance per se, this duality is only part of the target of the book. To reach an understanding of the subtleties involved in the duality, one has to take a journey through the properties of quantum integrable systems, building from the ground up the theory of exact S-matrices and familiarising oneself with the mathematical concept of a quantum group. This book therefore becomes an opportunity for a focused study of integrability in its wider breadth of interest, while always maintaining a clear ultimate purpose in mind: understanding the duality between bosons and fermions in $1 + 1$ dimensions. This should make going through the book from the point of view of the reader or early-career researcher a live enterprise, as opposed to a more passive learning exercise.

This book will start by describing the classical and the quantum Sine-Gordon model, in particular the quantum spectrum, S-matrices and underlying quantum-group symmetry, and the renormalisation properties. It will then proceed to discuss the Thirring model. This book will build the theory and at the same time will provide a significant number of examples. This book will then present the duality as originally stated by S. Coleman and refined in subsequent literature. It will also focus on a variety of tests of the duality. All the basic elements will be provided without relying on prerequisites beyond standard graduate-level quantum field theory textbooks. This book is disseminated with a series of exercises, some of which are explicitly solved and others which are more open-ended literature reviews.

This book originates from a series of lectures that were delivered at the Royal Institution of London in Fall 2022, for an audience of PhD students and early-career researchers of the London area, upon invitation by Bogdan Stefański Jr, Nadav Drukker and Yang-Hui He. While preparing the lectures, I immediately realised that the relevant information about this important chapter of $1 + 1$-dimensional mathematical physics was scattered over a vast amount of literature, with no resource making a systematic effort to collect, review, and contextualise the whole bulk of the material. I hope that this book will fill this gap and will become a valuable reference of where to find the literature organised and rationalised, most importantly with a consistency of notation and conventions. I also hope that this book might serve a purpose as a textbook for a PhD course or a graduate-level mini-module on the Sine-Gordon/Thirring duality. This is an extremely rich and important subject, but, to my knowledge, it is seldom taught at PhD schools. Moreover, while the matter is known to practitioners, senior colleagues might still find it of use to have a place where all the relevant information is concentrated, whether they intend to teach it or just refresh their knowledge of it.

Keeping in mind the perspective of an early-career researcher who might wish to study this book, I have favoured a format which keeps it rather short, where I have specifically avoided writing a long monograph. I find that a short series of focused lecture-style chapters keeps the attention of the young reader alive and does not intimidate the beginner—as opposed to being confronted with a weighty tome. The open-ended literature reviews that I suggest in various places throughout the book may constitute ideas for end-of-year projects, while the more standard exercises with solutions may make for ordinary problem sheets.

1.1 Prelude

Integrable systems (for a possible starting point, see [2–6]) have a long history. They stem from the longing for exact solutions to problems of interest in the natural sciences, starting with Newton's laws. The Kepler problem of planetary motion is an example of an integrable system which Newton himself did solve [7]. Such problems which admit an exact solution are rare. It was Liouville who attempted to perform a classification in the 1800s. Liouville's method applies to Hamiltonian systems and gives a general criterion to be able to completely solve a problem in principle by a reduction to quadratures, i.e., by reducing the solution to elementary integration and algebraic manipulations. A method to systematically obtain exact solutions was then developed by Gardner, Green, Kruskal and Miura in 1967, which was later dubbed the classical inverse scattering method. This method was first applied to the Korteweg–de Vries (KdV) equation in the context of fluid dynamics [8], see also [9, 10], and it allows one to find soliton-like solutions to nonlinear equations and to completely classify such solutions.

With the advent of quantum mechanics, the problem was posed of extending the notion of integrability to quantum mechanical systems. Since classical integrability is intimately connected with the Hamiltonian formulation, a natural generalisation to quantum mechanics can be found. It was, however, only with the Leningrad—St. Petersburg school [11, 12] that a quantum mechanical version of the inverse scattering method was constructed. In this method, a connection with the theory of Hopf algebras and quantum groups [13] was naturally established, setting the stage for the formulation of completely algebraic solutions of quantum integrable systems. In this way, both integrable quantum field theories [14] and spin-chains/vertex and lattice systems (cf [15]) are solved by one and the same set of tools.

Along the road it also became clear that *integrability* has a much more narrow and precise meaning than mere *exact solvability*. The set of integrable systems neither coincides with nor is a subset of the set of exactly solvable systems. There are, for instance, integrable systems which are not exactly solvable in practice (only in principle). If one considers that solving an integrable quantum field theory in finite volume requires a system of integral equations known as the thermodynamic Bethe ansatz to be solved [16], then one can see that finding an exact analytic solution is often (though not always) out of the question. Nevertheless, we understand the solution and can access it numerically, which is far more than can typically be claimed in a generic $3 + 1$-dimensional interacting quantum field theory. On the other hand, there are exactly solvable systems that are not integrable. One can think of simple problems with friction in one dimension, or of what is known as solvable chaos [17].

While the attribute of a system being solvable is in fact a specification of one's ability and/or calculational reach, integrability refers to a very specific set of properties, such as regular (quasiperiodic) behaviour and a large enough symmetry algebra to constrain the dynamics. Quantum mechanically, this means having a large amount of commuting charges and symmetries realised on the quantum spectrum. For such systems one has a toolbox of algebraic methods at one's disposal, which not only allows us to carry out the solution to a far greater extent than in generic non-integrable systems but also allows us to organise the solution and our understanding of it in such a way that we hopefully learn something universal in the process.

References

[1] Beisert N et al 2012 Review of AdS/CFT integrability: an overview Lett. Math. Phys. **99** 3–32

[2] Babelon O, Bernard D and Talon M 2003 Introduction to Classical Integrable Systems Cambridge Monographs on Mathematical Physics (Cambridge: Cambridge University Press)

[3] Faddeev L D and Takhtajan L A 1987 Hamiltonian Methods in the Theory of Solitons (Berlin: Springer)

[4] Novikov S, Manakov S V, Pitaevsky L P and Zakharov V E 1984 Theory of Solitons. The Inverse Scattering Method (Moscow: Academy of Sciences of USSR)

[5] Arutyunov G 2006 Student seminar: classical and quantum integrable systems Lecture Notes for Lectures Delivered at Utrecht University **20**

[6] Candu C and de Leeuw M 2013 Introduction to integrability Lectures delivered at ETH

[7] Speiser D 1996 The Kepler problem from Newton to Johann Bernoulli Arch. Hist. Exact Sci. **50** 103–16

[8] Gardner C S, Greene J M, Kruskal M D and Miura R M 1967 Method for solving the Korteweg-deVries equation Phys. Rev. Lett. **19** 1095–7

[9] Zakharov V E and Shabat A B 1974 A scheme for integrating the nonlinear equations of mathematical physics by the method of the inverse scattering problem. i Funktsional. Anal. i Prilozhen **8** 43–53

[10] Zakharov V E and Shabat A B 1979 Integration of nonlinear equations of mathematical physics by the method of inverse scattering. II Funct. Anal. Appl. **13** 166–74

[11] Sklyanin E K 1980 Quantum version of the method of inverse scattering problem Zap. Nauchn. Semin. **95** 55–128

[12] Faddeev L D and Zakharov V E 1971 Korteweg–de Vries equation: a completely integrable Hamiltonian system Funct. Anal. Appl. **5** 280–7

[13] Drinfeld V G 1986 Quantum groups Zap. Nauchn. Semin. **155** 18–49

[14] Zamolodchikov A B and Zamolodchikov A B 1979 Factorized S-matrices in two-dimensions as the exact solutions of certain relativistic quantum field models Annals Phys. **120** 253–91

[15] Baxter R J 1982 Exactly Solved Models in Statistical Mechanics (San Diego, CA: Academic Press Inc.)

[16] van Tongeren S J 2016 Introduction to the thermodynamic Bethe ansatz J. Phys. A **49** 323005

[17] Ulam S M 1947 On combination of stochastic and deterministic processes Am. Math. Soc. **53** 1120

[18] Coleman S R 1975 The Quantum Sine-Gordon Equation as the Massive Thirring Model Phys. Rev. D. **11** 2088

IOP Publishing

Integrability using the Sine-Gordon and Thirring Duality
An introductory course
Alessandro Torrielli

Chapter 2

Invitation to integrable quantum field theories

Quantum field theory is typically not solvable in closed form and represents a very complex and extremely accurate description of elementary particle physics. Integrable quantum field theories are sufficiently simplified and yet maintain a huge richness of interesting effects. They normally reside in a $1 + 1$-dimensional world and are characterised by exact S-matrices, which are calculable in closed form starting from fundamental physical requirements, such as unitarity, crossing symmetry and the location of bound-state singularities. Thanks to their vast degree of symmetry, such models are an ideal playground for training our understanding of nonpertubative quantisation. The theory of integrable systems also has a huge impact on contemporary advances in string theory [1].

For recent reviews and textbooks on the subject, the reader is invited to consult [2–5].

2.1 Classical integrability

In this section we follow [6].

Classically integrable field equations have the remarkable property that, despite being nonlinear, they can be treated with exact methods and their solutions retain certain features of linear systems. This is due to the existence of an infinite number of independent classical conservation laws, encoded in an object called the transfer matrix, which in turn is built out of the so-called monodromy matrix. The monodromy matrix is constructed using the Lax pair of the system. The Lax pair is a pair of matrices L, M, depending on the fields, the coordinates and an additional complex variable u called the *spectral parameter*, such that the equations of motion are equivalent to the system

$$\partial_t L - \partial_x M = [M, L],$$

(2.1.1)

which singles out a flat connection. We will see later an example of Lax pair when we will discuss the Sine-Gordon model.

doi:10.1088/978-0-7503-5899-6ch2

The idea of a Lax pair is in fact quite simple. It can be seen in the equation $\frac{d}{dt} A = [A, B]$, where A, B are any two matrices of the same dimension. The monomials $\mathrm{tr} A^n$ are all conserved, i.e. time independent, for any natural number n. This is easy to show due to the properties of the trace and the commutator. One can organise all the conserved quantities into a generating function $\mathrm{tr} \exp(A)$, which is effectively a prototype of transfer matrix. We will shortly see that this is quite a general procedure and one always obtains conservation laws from traces of a suitable monodromy matrix (the latter being an ordered exponential, as we will see).

The constraint (2.1.1) is literally the *integrability*—basically, consistency—condition (which is a necessary condition) for the solution of the system

$$\partial_x F = L F,$$
$$\partial_t F = M F, \tag{2.1.2}$$

F being any differentiable vector field function of (x, t). It is sufficient to differentiate the first equation with respect to time and the second equation with respect to space, and use the system again, to obtain the integrabilty condition. We call (2.1.2) the *auxiliary linear problem* associated with the integrable system.

In general, finding such a Lax-pair representation of the equations of motion, if it exists at all, is almost like an art, though some constructive methods have been developed (see [7]).

The all-important *monodromy matrix* is given by the path-ordered exponential (Wilson line)

$$T(u) = P \exp \int L \, dx. \tag{2.1.3}$$

We refer to [6] for details on how this quantity is used to produce, upon a suitable expansion in u, a tower of conserved quantities. It is easy to see that, if we call $T_{ab} = P \exp \int_a^b L dx$, then, using the Lax equation (2.1.1), one gets for $\partial_t T_{ab}$ the result

$$\int_a^b dx \, T_{xb} \, \partial_t L \, T_{ax} = \int_a^b dx \, T_{xb} \, (\partial_x M + [M, L]) \, T_{ax}$$
$$= \int_a^b dx \, \partial_x [T_{xb} M T_{ax}] = M(b) T_{ab} - T_{ab} M(a). \tag{2.1.4}$$

If we set periodic boundary conditions $M(a) = M(b)$, then the result equals $[M(a), T_{ab}]$. This means that

$$\partial_t \mathrm{tr} T_{ab} = 0, \tag{2.1.5}$$

since the trace of a commutator vanishes. The dependence on u has sometimes been left implicit in the above formulas, but it should appear everywhere. It is now simple to see that expanding $\operatorname{tr} T_{ab}$, for instance in powers of u, produces an infinite tower of (Taylor or Laurent) coefficients[1], which are all conserved by virtue of (2.1.5).

The issue of these charges being in involution, i.e. all Poisson-commuting with one another (a requirement of Liouville integrability), is also reviewed in [6]. We report here the best studied type of Poisson structure relevant to our discussion, i.e., the case of the *Sklyanin brackets*.

The Sklyanin brackets arise when the Poisson brackets between the different entries of the Lax matrix L are governed by a quantity r, like so:

$$\{L_1(x, t, u), L_2(y, t, u')\} = [r_{12}(u - u'), L_1(x, t, u) + L_2(y, t, u')]\,\delta(x - y). \quad (2.1.6)$$

The notation will always go like this: $A_1 \equiv A \otimes 1$, $A_2 \equiv 1 \otimes A$ (and similarly for more spaces), having introduced a tensor product of multiple spaces where the matrix A can reside. The quantity r is called the *classical r-matrix*, and we will request it not to depend itself on the fields in addition to satisfying

$$r_{12}(u - u') = -r_{21}(u' - u), \quad (2.1.7)$$

where $_{21}$ indicates permuting the two spaces: on a basis E_I of (Lie) algebra generators, we have $r_{12}(x) = r_{IJ}(x)E_I \otimes E_J$ and $r_{21}(x) = r_{IJ}(x)E_J \otimes E_I$. An object like r_{12} naturally lives in both spaces of the tensor product that we have introduced.

Under the assumption (2.1.6) and other ones that are very carefully surveyed in [8, 9], the Poisson brackets between the entries of the monodromy matrix can be recast in the form

$$\{T_1(u), T_2(u')\} = [r_{12}(u - u'), T_1(u)T_2(u')]. \quad (2.1.8)$$

If we can arrive at the set of relations (2.1.8), we can conclude that the system is integrable in the sense of Liouville. In fact, by tracing the relation (2.1.8) using the natural trace operation $\operatorname{tr}_1 \otimes \operatorname{tr}_2$, we get

$$\{\operatorname{tr} T(u), \operatorname{tr} T(u')\} = 0, \quad (2.1.9)$$

having also used the cyclicity property of the trace. The quantity $\operatorname{tr} T(u)$ works as a generating function of the conserved charges: if we expand (2.1.9) in series, then we have immediately

$$\operatorname{tr} T(u) = \sum_{n \geqslant 0} Q_n u^n, \qquad \{Q_n, Q_m\} = 0 \quad \forall\, m, n \geqslant 0, \quad (2.1.10)$$

assuming analyticity in u.

We can say that the entries of the monodromy matrix are the variables where the integrability of the system is better manifested. They will naturally also be the best variables—when suitably interpreted quantum mechanically—to quantise the system while preserving integrability [10], as we will see in explicit examples throughout this book.

---

[1] Depending on the problem, other types of expansion might be more natural.

Having written the relations (2.1.8), one might wonder whether one can explicitly verify that they are consistent with the Jacobi identity. A sufficient condition for this to happen, under the stated assumptions of (2.1.7) and r being field-independent, is

$$[r_{12}(u_1 - u_2), r_{13}(u_1 - u_3)] + [r_{12}(u_1 - u_2), r_{23}(u_2 - u_3)]$$
$$+ [r_{13}(u_1 - u_3), r_{23}(u_2 - u_3)] = 0, \qquad (2.1.11)$$

which needs to be satisfied in the physical representation at hand. The relation (2.1.11) goes under the name of *classical Yang–Baxter equation*.

The type of Poisson brackets (2.1.6) is referred to as *ultralocal* because the right-hand side only contains the Dirac delta and not derivatives thereof. If derivatives of the delta function are present, then we have non-ultralocal Poisson brackets [8, 9]. In the non-ultralocal case things are much harder and the quantum inverse scattering method is traditionally much more involved—for recent developments, the reader is for instance referred to [11, 12]. We will see later an example of non-ultralocal Poisson brackets—see section 9.4. For the case of the Sine-Gordon model, we refer the reader to [13, 14].

Let us remark that the spectral parameter is not a dynamical variable of the theory, in that it does not couple to the fields and is completely absent from any on-shell data. It is also not uniquely defined because one can introduce reparameterisations[2]. The variable u can be thought of as an organising parameter for the generating function of the conserved charges. The very fact that it exists, meaning that the Lax pair permits the freedom of an arbitrary complex parameter, is probably a very deep reflection of integrability itself. Ultimately, it allows for the powerful tools of complex analysis to enter the game, and it is known that integrable systems owe much of their striking features to a profound complex-analytic structure.

2.2 Exact *S*-matrices

In this section, we follow [15, 16].

Quantising a classically integrable system in a way that preserves the conservation laws results in a quantum integrable system[3]. The presence of infinitely many conservation laws constrains the dynamics to a point where the scattering of quantum particles is reduced to the following:

1. No particle production/annihilation is admitted;
2. The initial and final sets of momenta are preserved—momenta are only reshuffled;

---

[2] The Lax pair itself is also by no means unique, partly because of a gauge freedom, as described in [6].

[3] There are of course quantum integrable systems with no classical analogue—one class of such examples being spin-chains. The terminology 'spin-chain' simply means a one-dimensional array of distinguishable sites, each carrying a representation of some symmetry algebra (the 'spin' part of the name descends from the prototypical example where such symmetry algebra is $\mathfrak{su}(2)$). From spin-chains, one can obtain a classical integrable system by taking a continuous limit. Such classical systems then admit the original spin-chain as a (lattice) quantisation (discretisation).

3. *Factorisation*: The N-body S-matrix factorises in a sequence of two-body processes—the ordering of the sequence does not matter by virtue of the Yang–Baxter equation (see later).

It is important to remark that these properties refer to the *full* (exact) scattering amplitude between particles of the *exact* quantum spectrum, while individual Feynman graphs do of course display production or annihilation of perturbative excitations because the Lagrangian description has typically got interaction vertices. The resummation of the perturbative series is, however, bound by symmetries to respect the above constraints.

All we need to study is therefore the two-body S-matrix for all the particles in the spectrum. The particles have internal degrees of freedom (isospin, flavour, etc), which means that typically the two scattering states, labelled 1 and 2, come in multiplets carrying representations, of dimension d_1 and d_2, respectively, of the symmetry algebra of the problem. The S-matrix is therefore a $(d_1 d_2) \times (d_1 d_2)$ matrix whose entries depend on the momenta p_1 and p_2 of the two scattering particles. Such a matrix, which acts on the tensor product of the two vector spaces $V_1 \otimes V_2$ representing the internal degrees of freedom of each scattering particle 1 and 2, can still be very complicated.

If we focus on relativistic theories, we can parameterise the energy and momentum of particles in $1 + 1$ dimensions as $E_i = m_i \cosh \theta_i$ and $p_i = m_i \sinh \theta_i$, where $\theta_i \in \mathbb{R}$ is the *rapidity* and $i = 1, 2$ labels the particle. Notice that we have explicitly assumed that *the masses are nonzero*—otherwise an entire new story will have to apply (see much later on in this book). Since a Lorentz boost shifts all of the rapidities by one and the same constant, relativistic invariance then dictates that the S-matrix only depends on the difference of the rapidities of the scattering particles:

$$S_{12} = S_{12}(\theta_1 - \theta_2) = S_{12}(\theta), \qquad \theta = \theta_1 - \theta_2. \tag{2.2.1}$$

The 'axioms' (the word to be taken with caution, as we will later comment) of integrable scattering allow one to *construct* the S-matrix from first principles, knowing the symmetries and the particle content of the model. Such axioms are *schematically*[4] given by

1. Braiding unitarity: $S_{12}(\theta)S_{21}(-\theta) = \mathbb{1} \otimes \mathbb{1}$, where $\mathbb{1}$ is the identity matrix in the internal space of each particle—see figure 2.1;
2. Crossing symmetry: $S_{12}(\theta) = S_{\bar{2}1}(i\pi - \theta)$, where $\bar{2}$ is the antiparticle of 2—see figure 2.2;
3. Yang–Baxter equation: $S_{12}(\theta_1 - \theta_2)S_{13}(\theta_1 - \theta_3)S_{23}(\theta_2 - \theta_3) = S_{23}(\theta_2 - \theta_3)$ $S_{13}(\theta_1 - \theta_3)S_{12}(\theta_1 - \theta_2)$, see figure 2.3;
4. Physical unitarity: the S-matrix is a unitary matrix for real values of the rapidities;

---

[4]The language of Hopf algebras and their representations, which we will see later on, allows for a mathematical translation of these axioms in a way that can be implemented using standard representation-theoretical tools.

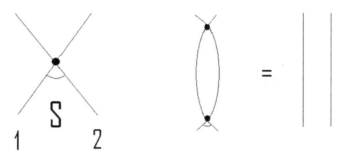

Figure 2.1. The two-body S-matrix and its braiding unitarity property.

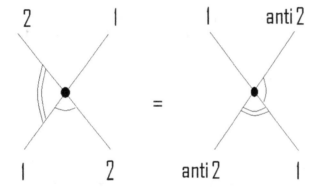

Figure 2.2. Crossing symmetry: the image on the left is seen from the perspective of the direct (s-)channel, while the image on the right is seen from the perspective of the crossed (t-)channel (rotating by 90°).

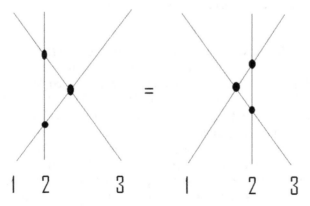

Figure 2.3. The Yang–Baxter equation, which says that the ordering of the sequence of two-body scatterings does not matter.

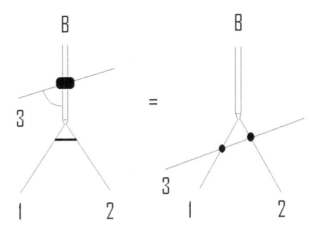

Figure 2.4. The bootstrap principle: the (hyperbolic) angle denoted by a thick line is the value at which **B** appears as a simple pole in S_{12}. All the angles in the image are then fixed once the scattering angle of 3 with **B** is specified.

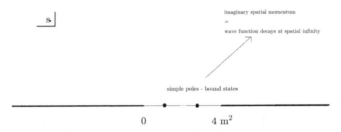

Figure 2.5. The analytic structure of the S-matrix in a typical isospin or flavour channel expressed in the Mandelstam variable s (centre of mass energy). If the mass of the two scattering particles is the same, then $s = 2m^2(1 + \cosh\theta)$. The branch point starting at $4m^2$ (s-channel cut) corresponds to the threshold for two particles going on-shell, while the cut starting from 0 is the corresponding t-channel cut, present because of crossing symmetry.

5. Bootstrap: if the particles 1 and 2 can form a bound state **B**, then the S-matrix for the process where a third particle 3 is scattering against **B** is given by $S_{3B}(\theta) = S_{32}(\theta + ix)S_{31}(\theta + iy)$, where x and y are calculated knowing the (complex) rapidity difference $\theta_1 - \theta_2$ at which the S-matrix S_{12} has the corresponding bound-state pole—see figure 2.4. The reader is also invited to consult [17];

6. As always in scattering theory, it is useful to complexify all the momenta and exploit the power of analyticity; the S-matrix is a meromorphic (matrix-valued) function in the complex θ-plane, with possible poles and zeros. Bound states correspond to simple poles in specific (isospin, colour, flavour) channels on the imaginary segment Im $\theta \in (0, \pi)$, Re $\theta = 0$; the analytic structure in the Mandelstam plane also has cuts, and is depicted in figure 2.5. The entire visible portion of the Mandelstam plane is mapped to the so-called *physical strip* Im $\theta \in (0, \pi)$ in the complex θ-plane.

7. If the theory is parity-invariant, then the S-matrix is required to satisfy a parity condition—see the discussion in [16, 18]. We also refer to [18] for more insights on the transformation properties under time-reversal and for a more general presentation that does not make use of relativistic invariance.

We have not enumerated as an 'axiom' the condition that the S-matrix must respect the exact quantum symmetries of the model. We will see later how to rephrase this in a Hopf-algebraic language, in a way that will make it amenable to standard methods of representation theory. This is in practice almost always the first step in the process of determining the exact S-matrix of the model. It is true that one does not always immediately (and sometimes, ever) know all the quantum symmetries of the system, but they are not all necessary. On many occasions it is sufficient to start with few obvious structural ones (which may for instance set to zero some S-matrix entries by some obvious selection rule) and let the 'axioms' above do the rest. With the knowledge of the exact S-matrix, it is later possible to discover *a posteriori* what the hidden symmetries were (see our later discussion on the so-called *universal R-matrix*).

Notice that in $1 + 1$-dimensional integrable scattering (with all the particles having the same mass) one has (since $p_3 = p_1$ and $p_2 = p_4$, and given that we use the signature with -1 along the time direction, such that $p_i^2 = -m^2$)

$$s = -(p_1 + p_2)^2, \quad t = -(p_1 - p_4)^2 = -(p_1 - p_2)^2 = 4m^2 - s, \quad u = -(p_1 - p_3)^2 = 0.$$

To give an example of bootstrap, one can take the S-matrix of the Lee–Yang model, which involves a single scalar particle (no matrix structure). The Lee–Yang particle is a bound state of itself:

$$S_{LY}(\theta) = \frac{\sinh \theta + i \sin \dfrac{\pi}{3}}{\sinh \theta - i \sin \dfrac{\pi}{3}}, \qquad S_{LY}(\theta) = S_{LY}\left(\theta + i\frac{\pi}{3}\right) S_{LY}\left(\theta - i\frac{\pi}{3}\right), \qquad (2.2.2)$$

where the simple pole at $\theta = \frac{2}{3}i\pi$ is the bound state, which is the very same particle as 1 and 2. This is also confirmed by the calculation of the Mandelstam invariant: at the bound-state pole we have in fact $s = -(p_1 + p_2)^2 = -2m^2(1 + \cosh \theta) \rightarrow -2m^2(1 + \cosh \frac{2}{3}i\pi) = -m^2$. The bootstrap condition in (2.2.2) has an easily workable graphical construction—see figure 2.6. For a thorough discussion of the integrable S-matrix for a particle that is a bound state of itself, the reader is invited to consult chapter 18 in [19]. To conclude this small detour on the Lee–Yang model, following section 2 of [20], one can prove that the residue of S_{LY} at the bound-state pole has the 'wrong' sign. This signals that unitarity ultimately breaks down—the way in which it does so being rather subtle, as explained in chapter 18 of [19]. The reader can refer to the supplement for an explicit demonstration.

At this stage we can provide a justification as to why we have restricted ourselves to $1 + 1$ dimensions. Quantum integrable systems are characterised by a vast number of charges that commute with the quantum Hamiltonian and amongst

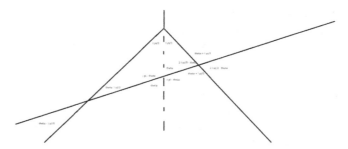

Figure 2.6. The bootstrap principle (see the diagram after the equal sign in figure 2.4) for the Lee–Yang model is effectively based on simple 'angular' relations.

themselves. In higher dimensions, such as $3 + 1$, the presence of these charges would typically imply a trivial scattering matrix via the Coleman–Mandula theorem [21]. In $1 + 1$ dimensions, such a theorem is avoided (even without invoking supersymmetry). Very roughly, the reason is that in one space and one time dimension, it is quite hard to deflect two particle trajectories away from each other—see [15, 16]. No matter how many charges (symmetry transformations) one might apply, the two particles will almost invariably meet. What the conservation laws will produce is a great number of constraints on the functional form of the (nontrivial) scattering matrix. Most of these constraints are then subsumed into the scattering 'axioms'.

Let us also mention that the 'axioms' reported above are a somewhat reductive list. They can be extended to nonrelativistic models [18] and ultimately they can be formalised into the language of Hopf algebras [16, 22, 23]—see also section 3.6. Hopf algebras no longer essentially distinguish whether one is dealing with relativistic field theories, or quantum-mechanical spin-chains, or Galilean models. A comment is, however, necessary at this stage. Even remaining within the domain of relativistic models, the 'axioms' as we have presented them here are ultimately not to be considered with mathematical rigidity[5]. There are integrable models where not all the 'axioms' are satisfied, yet such models may still describe interesting physics. While some relations must *always* be there (such as the Yang–Baxter equation, or the conditions stemming from a local relativistic formulation such as physical unitarity and crossing symmetry), others might be relaxed if the physics so requires —these models might, for instance, be described by mathematical constructions even more general than Hopf algebras. The relations that always appear typically descend from the principles of local relativistic quantum field theory, with an added bonus: integrability makes the writing of those principles more easily workable. Effectively this is because, as we will see much later, in theories with such a constrained scattering one has control on how to write the resolution of the identity onto the whole spectrum exactly and compactly. In many practical applications, the spirit in

---

[5] This is not to say that there do not exist theorems about integrable scattering which one can rigorously prove [15, 16]. For recent work on axiomatic quantum integrability, the reader is referred for instance to [24, 25].

which the construction should be taken remains to be best described as follows. Given any specific theory, a stubborn go at perturbation theory would ultimately reveal which properties are verified and which are not—but the 'axioms' shortcut the whole process by giving us something to aim for and by asking us to take a little (and rather well-informed) leap of faith. The reward is deriving exact expressions, to later be perturbatively tested to our heart's content.

This S-matrix programme (which at one point in history held the promise of becoming the dominant paradigm for particle physics [26]) finds a life of its own in 1 + 1-dimensional physics. The idea is, therefore, to derive the S-matrix exactly from symmetries (we will say more on this in section 3.6) and to impose the above 'axioms'. Parameters that are left undetermined are then matched to perturbation theory or semiclassics. One important ambiguity left over by this approach is the one represented by the so-called Castillejo–Dalitz–Dyson (CDD) [27] factors: these are scalar factors by which any S-matrix can be multiplied while still preserving its basic properties. One particular form which CDD factor can have is

$$S_{CDD}(\theta) = \prod_{i=1}^{M} \frac{\sinh \theta + i \sin \alpha_i}{\sinh \theta - i \sin \alpha_i}, \qquad \alpha_i \in \mathbb{R} \; \forall \; i = 1, \ldots , M, \qquad (2.2.3)$$

in a relativistic-invariant theory. One can easily see that they are pure phases for real rapidities and that

$$S_{CDD}(-\theta) = S_{CDD}^{-1}(\theta), \qquad S_{CDD}(i\pi - \theta) = S(\theta), \qquad (2.2.4)$$

hence they neither spoil physical nor braiding unitarity, nor crossing. What can change, however, is the pole structure, hence the physics is in principle dramatically altered by the presence of a CDD factor. Needless to say, these factors can be used to adjust a proposed solution in such a way that it reflects the expected spectrum of bound states (argued, for instance, using a semiclassical or a perturbative reasoning).

Project *[6 weeks' work]: Taking the moves from* [28, 29], *explore the literature on the so-called $T\bar{T}$ deformation.*

References

[1] Beisert N *et al* 2012 Review of AdS/CFT integrability: an overview *Lett. Math. Phys.* **99** 3–32

[2] Bombardelli D, Cagnazzo A, Frassek R, Levkovich-Maslyuk F, Loebbert F, Negro S, Szécsényi I M, Sfondrini A, van Tongeren S J and Torrielli A 2016 An integrability primer for the gauge-gravity correspondence: an introduction *J. Phys.* A **49** 320301

[3] Retore A L 2022 Introduction to classical and quantum integrability *J. Phys.* A **55** 173001

[4] Driezen S 2022 Modave lectures on classical integrability in 2D field theories *PoS* **Modave2021:002**

[5] Arutyunov G 2019 *Elements of Classical and Quantum Integrable Systems* (Berlin: Springer)

[6] Torrielli A 2016 Lectures on classical integrability *J. Phys.* A **49** 323001

[7] Babelon O, Bernard D and Talon M 2003 *Introduction to Classical Integrable Systems* (Cambridge: Cambridge University Press) Cambridge Monographs on Mathematical Physics.

[8] Maillet J M 1986 Hamiltonian structures for integrable classical theories from graded Kac-Moody algebras *Phys. Lett.* B **167** 401–5

[9] Maillet J M 1986 New integrable canonical structures in two-dimensional models *Nucl. Phys.* B **269** 54–76

[10] Sklyanin E K 1980 Quantum version of the method of inverse scattering problem *Zap. Nauchn. Semin.* **95** 55–128

[11] Delduc F, Magro M and Vicedo B 2012 Alleviating the non-ultralocality of coset sigma models through a generalized Faddeev-Reshetikhin procedure *JHEP* **08** 019

[12] Vicedo B 2020 On integrable field theories as dihedral affine gaudin models *Int. Math. Res. Not.* **2020** 4513–601

[13] Sklyanin E K, Takhtadzhyan L A and Faddeev L D 1979 Quantum inverse problem method. i *Teor. Mat. Fiz.* **40** 194–220

[14] Kundu A and Ghosh S 1988 Soliton and breather states of the quantum sine-Gordon model in light cone coordinates through the exact qist method *J. Phys. A Math. Gen.* **21** 3951

[15] Dorey P 1996 *Eotvos Summer School in Physics: Conformal Field Theories and Integrable Models* (Berlin: Springer) 85–125 pp

[16] Bombardelli D 2016 S-matrices and integrability *J. Phys.* A **49** 323003

[17] Karowski M 1979 On the bound state problem in $(1 + 1)$-dimensional field theories *Nucl. Phys.* B **153** 244–52

[18] Arutyunov G and Frolov S 2009 Foundations of the $AdS_5 \times S^5$ superstring. Part I *J. Phys.* A **42** 254003

[19] Mussardo G 2020 *Statistical Field Theory.* Oxford Graduate Texts **3** (Oxford: Oxford University Press)

[20] Paulos M F, Penedones J, Toledo J, van Rees B C and Vieira P 2017 The S-matrix bootstrap II: two dimensional amplitudes *JHEP* **11** 143

[21] Coleman S R and Mandula J 1967 All possible symmetries of the S matrix *Phys. Rev.* **159** 1251–6

[22] Loebbert F 2016 Lectures on Yangian symmetry *J. Phys.* A **49** 323002

[23] Spill F 2007 Hopf algebras in the AdS/CFT correspondence *Master's thesis* Humboldt U., Berlin

[24] Lechner G 2008 Construction of quantum field theories with factorizing s-matrices *Commun. Math. Phys.* **277** 821–60

[25] Bostelmann H, Cadamuro D and Fewster C J 2013 Quantum energy inequality for the massive ising model *Phys. Rev.* D **88** 025019

[26] Eden R J, Landshoff P V, Olive D I and Polkinghorne J C 1966 *The analytic S-matrix* (Cambridge: Cambridge University Press)

[27] Castillejo L, Dalitz R H and Dyson F J 1956 Low's scattering equation for the charged and neutral scalar theories *Phys. Rev.* **101** 453–8

[28] Smirnov F A and Zamolodchikov A B 2017 On space of integrable quantum field theories *Nucl. Phys.* B **915** 363–83

[29] Cavaglià A, Negro S, Szécsényi I M and Tateo R 2016 $T\bar{T}$-deformed 2D quantum field theories *JHEP* **10** 112

IOP Publishing

Integrability using the Sine-Gordon and Thirring Duality
An introductory course
Alessandro Torrielli

Chapter 3

The Sine-Gordon model

3.1 A very special theory

The Sine-Gordon model is certainly one of the most famous integrable field theories of all. Let us begin with a very nice intuitive argument from [1] as to why this particular model should be integrable. If one starts with the familiar scalar theory in $1 + 1$ dimensions

$$-\frac{1}{2}\partial_\mu \phi \partial^\mu \phi - \frac{m^2}{2}\phi^2 - \frac{\lambda}{4!}\phi^4, \tag{3.1.1}$$

then one can compute the tree-level production amplitude $2 \to 4$. Such an amplitude turns out to be constant and it can be cancelled by adding a term $-\frac{\lambda^2}{6!m^2}\phi^6$ to the original Lagrangian. At this point one looks at the $2 \to 6$ amplitude in this new theory and discovers that this amplitude too can be cancelled by adding a specific ϕ^8 term. This procedure can continue until one has by hand eliminated all of the possible tree-level production amplitudes; of course, at the cost of having added an infinite number of terms to the original Lagrangian. Those infinitely many terms form the Taylor series of $-\frac{m^2}{\beta^2}(\cosh \beta\phi - 1)$, where $\beta^2 = \frac{\lambda}{m^2}$. The Sine-Gordon theory is then recovered by continuing $\beta \to i\beta$. In a sense, the infinitely many symmetries of integrability tie all of the interaction monomials into being the Taylor series of a very specific function. Along the same lines, one sees that quantum integrability is a symmetry that binds the renormalisation of all the monomials to proceed in unison.

Page 4 in [1] explains why the argument, which we have just sketched above, does *not* imply the vanishing of the $3 \to 3$ amplitude (which is in fact quite importantly nonzero and constitutes the basis for writing the Yang–Baxter equation). The superficial expectation that the $3 \to 3$ amplitude be obtained by analytically continuing the $2 \to 4$ amplitude fails in two dimensions for an integrable kinematics, due to a subtle resolution of singularities in the Feynman diagrams [1].

doi:10.1088/978-0-7503-5899-6ch3

These arguments of cancellation of production/annihilation amplitudes also extend to the loop level as well, but it is easier to resort to other than diagrammatic methods [2–12]. The powerful tools of the quantum inverse scattering method can then be employed. We shall work throughout these notes in particular conventions that should align, as much as and wherever possible, with Coleman [13]. Ultimately, the thermodynamic Bethe ansatz (TBA) [14] furnishes the complete description of the spectrum of the model compactified on a spatial circle. The reader is referred to [15–17], where the emphasis is put on the method provided by the Destri–de Vega nonlinear integral equation [18], and to [19–21]—see also [22] for a nice recent summary.

We particularly recommend the excellent review [23] and thank Zoltan Bajnok for several illuminating discussions.

3.2 Classical aspects

The classical Sine-Gordon Lagrangian is given by

$$L_{SG} = -\frac{1}{2}\partial_\mu\phi\partial^\mu\phi + \frac{m^2}{\beta^2}(\cos\beta\phi - 1), \qquad m, \beta \in \mathbb{R}, \tag{3.2.1}$$

where $\phi = \phi(x, t)$ is a scalar (more accurately, it is taken to be a pseudo-scalar [24]) field in (flat) $1 + 1$ dimensions. We remind the reader that we work in the signature with time having the minus sign. We can assume $\beta > 0$ because really only β^2 matters to the Lagrangian. The -1 shift sets to zero the energy of the trivial vacuum $\phi = 0$. Both the field φ and the parameter β have engineering mass-dimension 0, while m has engineering mass-dimension 1.

The equations of motion (giving the name to the model) are

$$\partial_t^2\phi - \partial_x^2\phi + \frac{m^2}{\beta}\sin\beta\phi = 0 \tag{3.2.2}$$

and tend to those of a free massive Klein–Gordon field of mass m as $\beta \to 0$. The classical Lagrangian expands as

$$L_{SG} = -\frac{1}{2}\partial_\mu\phi\partial^\mu\phi - \frac{1}{2}m^2\phi^2 + \frac{1}{24}m^2\beta^2\phi^4 + \cdots \tag{3.2.3}$$

Truncating at the order ϕ^4 now clearly does not produce a potential which is bounded below, but we do not worry about this because we will be interested in the whole (integrable) potential, which must include all the higher order corrections (and is clearly bounded).

3.2.1 Solitons

The solutions corresponding to solitons and antisolitons are obtained by imposing the asymptotic conditions

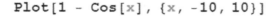

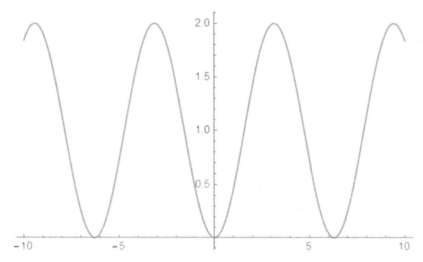

Figure 3.1. The Sine-Gordon potential in suitable units

$$\phi \to 0 \quad x \to -\infty, \qquad \phi \to \frac{2\pi}{\beta} \quad x \to \infty, \qquad (3.2.4)$$

or

$$\phi \to \frac{2\pi}{\beta} \quad x \to -\infty, \qquad \phi \to 0 \quad x \to \infty, \qquad (3.2.5)$$

for all t, where (3.2.4) is for the soliton and (3.2.5) for the antisoliton. There is a *topological* charge associated with these two solutions, given by

$$q_0 = \frac{\beta}{2\pi} \int_{-\infty}^{\infty} dx\, \phi' = \frac{\beta}{2\pi}[\phi(x \to \infty) - \phi(x \to -\infty)] = \pm 1, \qquad (3.2.6)$$

where we have used the asymptotic conditions. Because of the shape of the profile, these solutions are also called *kink* and *antikink*, respectively. Kink solutions exist because the classical potential has distinct minima—see figure 3.1, and the kink/antikink interpolate between the minimum at $\phi = 0$ and the two adjacent minima:

$$V(\phi) = \frac{m^2}{\beta^2}(1 - \cos \beta\phi) \quad \longrightarrow \quad \phi_{min} = 2n\frac{\pi}{\beta}, \quad n \in \mathbb{Z}. \qquad (3.2.7)$$

One can verify by brute force, for instance, that (for $m = 1$) the two static profiles

$$\phi = \frac{4}{\beta}\arctan e^{\pm(x-a)}, \qquad a \in \mathbb{R}, \qquad (3.2.8)$$

satisfy the equations of motion with the asymptotes (3.2.4) and (3.2.5), respectively —see figure 3.2. The parameter a is the centre of the kink.

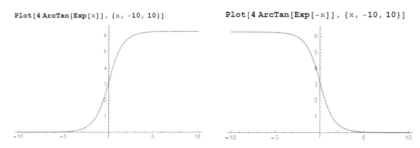

Figure 3.2. The Sine-Gordon static kink and, respectively, antikink, for $a = 0$ and in suitable units.

Notice that, since the Lagrangian is invariant under $\phi \rightarrow -\phi$, solutions that are not invariant under this map come in pairs $(\phi, -\phi)$. One can check that

$$-\phi_{kink} = \phi_{antikink} - \frac{2\pi}{\beta}, \qquad (3.2.9)$$

hence $-\phi_{kink}$ is also a solution (equal to an antikink pushed down by an allowed shift-symmetry of the action). This is very clearly visible from the plots in figure 3.2. Notice also that, the equation of motion being nonlinear, we cannot make linear combinations that are eigenstates of the transformation $\phi \rightarrow -\phi$. Nevertheless, integrability still allows for a certain notion of 'superposition' of (anti)kink solutions (see later).

A nonzero velocity can be given to the solutions [25]:

$$\phi(x, t) = \frac{4m}{\beta}\arctan\exp\left(\pm m\frac{x - a - vt}{\sqrt{1 - v^2}}\right), \qquad a \in \mathbb{R}, \qquad v \in (-1, 1). \ (3.2.10)$$

We can see the typical feature of integrable solitons to rigidly translate with their shape unmodified. Notice a very interesting feature of the solution: it has a particle-like behaviour [25]. Notice furthermore the Lorentz contraction factor, which guarantees that the scalar profile appears as static $\phi_{rest}(x_{rest}, t_{rest}) = \frac{4m}{\beta}\arctan\exp x_{rest} = \phi(x, t)$ in the rest-frame of the soliton itself (we are working in natural units where $c = 1$).

The charge q_0 (3.2.6) corresponds to the current $j^\mu = \epsilon^{\mu\nu}\partial_\nu\phi$, which is conserved by virtue of the commutativity of partial derivatives. Normally, such a charge would vanish for normalisable solutions, but in this case it acquires a topological meaning —see pages 18–9 of [26].

The attribute of *topological* for the charge q_0 stems from the fact that the field $\beta\phi$ can effectively be thought of as a (target space) angle. The soliton and antisoliton profiles are contained within the fundamental region $[0, 2\pi]$, and describe this angle winding the S^1 target space once in the positive, resp. negative direction, as x goes from $-\infty$ to $+\infty$. The conserved charge q_0 is just the winding number of the field regarded, at any fixed time, as a map $\beta\phi: \mathbb{R}^{1,1} \rightarrow S^1$. The conservation of this number and its additivity as a charge acquire therefore a topological nature [26].

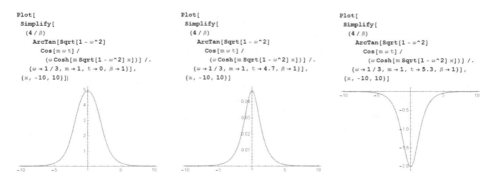

Figure 3.3. The non-translating Sine-Gordon breather in suitable units and for a chosen value of ω, as time evolves (snapshots at three subsequent times: $t = 0$, $t = 4.7$ and $t = 5.3$, respectively). Please notice the zoom on the vertical axis.

The (non-translating) *breather* is another notable solution of the equation of motion, given by

$$\phi = \frac{4}{\beta} \arctan \frac{\sqrt{1 - \omega^2} \cos(m\omega t)}{\omega \cosh(m\sqrt{1 - \omega^2}x)}, \qquad \omega \in [0, 1]. \tag{3.2.11}$$

The breather oscillates up and down on the spot (because we have set the velocity of translation to zero)—see figure 3.3. The breather is, therefore, an 'eigenstate' of the map $\phi \rightarrow -\phi$, in the sense that $-\phi_{breather}$ is a breather oscillating in phase opposition with respect to the original breather (still a solution of course, and, although technically a different solution, for all practical purposes considered as 'the same solution').

An important remark is that two classical kinks (solitons) repel each other, as can be ascertained from looking at the time evolution of the two-kink solution[1]. The same happens to two antikinks (antisolitons). This will be important in the following, because it will be connected with the fact that the quantised soliton and antisoliton will behave like fermions[2].

Exercise [30 minutes' work using Mathematica]: verify that the soliton, antisoliton and breather profiles have a finite energy $E = \frac{1}{2}\dot{\phi}^2 + \frac{1}{2}\phi'^2 + \frac{m^2}{\beta^2}(1 - \cos \beta\phi)$, *in the case* $\beta = m = 1$, $v = a = 0$, $\omega = \frac{1}{3}$.

As described in [27], integrability allows for a systematic way of finding solutions by means of the classical inverse scattering method. One can then obtain multiple travelling profiles that scatter, preserving their individual identity. This means that the set of stable classical solutions, in addition to the infinitely many 'vacua' (where

---

[1] See for instance *https://people.maths.ox.ac.uk/trefethen/pdectb/sine-gordon2.pdf* (accessed on 3 March 2023) for a discussion on this point.
[2] Subtle issues related to this identification will be mentioned in the Conclusions.

the field is constant and equal to one of the minima of the potential), consist of 'superpositions' (not in the linear sense) of arbitrarily many elementary solutions of the types described above. Each of these profiles are travelling along the real line and are scattering (i.e., classically, spend a certain amount of time-delay when passing one another) maintaining their individuality of profiles (akin to multiparticle states with definite momenta on a line). The reader can consult for instance [28], page 587, for a clear demonstration of this. This picture finds a clear correspondence in the factorised scattering theory of the quantised version of these elementary profiles, i.e., the quantum particles of the spectrum.

3.2.2 Lax pair and classical inverse scattering

The Lax pair for the Sine-Gordon equation is given by (see for example [27])

$$
L(u) = i \begin{pmatrix} \dfrac{\beta}{4}\dot{\phi} & \dfrac{m}{4}ue^{i\frac{\beta}{2}\phi} - \dfrac{m}{4u}e^{-i\frac{\beta}{2}\phi} \\[2ex] \dfrac{m}{4}ue^{-i\frac{\beta}{2}\phi} - \dfrac{m}{4u}e^{i\frac{\beta}{2}\phi} & -\dfrac{\beta}{4}\dot{\phi} \end{pmatrix},
$$

$$
M(u) = i \begin{pmatrix} \dfrac{\beta}{4}\phi' & -\dfrac{m}{4}ue^{i\frac{\beta}{2}\phi} - \dfrac{m}{4u}e^{-i\frac{\beta}{2}\phi} \\[2ex] -\dfrac{m}{4}ue^{-i\frac{\beta}{2}\phi} - \dfrac{m}{4u}e^{i\frac{\beta}{2}\phi} & -\dfrac{\beta}{4}\phi' \end{pmatrix},
$$

(3.2.12)

where $\dot{\phi} = \partial_t\phi$ and $\phi' = \partial_x\phi$. The spectral parameter u is an arbitrary complex variable, as discussed at the beginning of the book.

Exercise [20 minutes' work using Mathematica]: Verify that (3.2.12) is a Lax pair for the Sine-Gordon equation, and observe first-hand how the spectral parameter decouples from the on-shell data.

In [27] one can find a description of the classical inverse scattering method to obtain solutions of the Sine-Gordon equation, based on the Gelfand–Levitan–Marchenko equation—whose starting point is the Lax pair. One then recovers in particular the soliton and antisoliton profiles described in the previous section by specifying a certain set of *classical scattering data*—which are not the quantum scattering matrices, but rather encode the spectral decomposition of a specific differential operator associated with the Lax pair.

Lax pairs do in fact belong to Lie algebras or deformations thereof [27]—in the case of Sine-Gordon the $2 \cdot 2$ matrices L, M displayed above are antiHermitian. This has a connection with the quantum-group symmetry of the quantised version of the model—in this case the affine version of $U_q(\mathfrak{su}(2))$, which we will describe later on. Typically, the quantum group reduces to the classical Lie algebra when a suitable parameter (playing the role of $\hbar$) is sent to 0.

We conclude this survey of the classical properties of the Sine-Gordon solitons by mentioning that the multi-soliton solutions of the Sine-Gordon equation are connected with the particle dynamics described by the so-called *Ruijsenaars–Schneider* model—see for instance [29, 30].

3.3 Quantum aspects

Perturbatively one can establish that the dimensionless coupling β is not renormalised, while the dimensional coupling m^2 is. This statement, which is based on Coleman's approach [13], is only valid for $\beta^2 < 8\pi$. For $\beta^2 \geqslant 8\pi$, an alternative renormalisation approach than Coleman's is required, which involves a running of the coupling β and a wave-function renormalisation. For reasons that go beyond the scope of this book, this alternative renormalisation approach is also more desirable in a good portion of the $\beta^2 < 8\pi$ region of the space of the parameters of the model. We will restrict in this book exclusively to $\beta^2 < 8\pi$ and to Coleman's approach. The reader needs to bear in mind that as β^2 grows closer and closer to 8π, the region of the space of the parameters of the model where the alternative renormalisation approach is instead more desirable tends to take over more and more, see figure 1 in [31]. We recommend the reader to consult [31] and references therein, where an in-depth description of these issues can be found. We also refer to [32, 33]. We thank Benoit Vicedo for extensive discussions and email exchanges on this aspect.

The Sine-Gordon Lagrangian is superficially super-renormalisable for $0 < \beta^2 < 8\pi$ and non-renormalisable for $\beta^2 > 8\pi$. One can see this by looking at the scaling dimension of the cosine term regarded as a perturbation of a free boson (see further down in this section). All the ultraviolet (UV) divergences in the region $0 < \beta^2 < 8\pi$ can be removed and the dimensional coupling m^2 gets multiplicatively renormalised. In the subregion $0 < \beta^2 < 4\pi$, normal ordering is sufficient to remove all divergencies—provided, as we have alluded to, that one multiplicatively renormalises the bare mass parameter m^2. For $4\pi \leqslant \beta^2 < 8\pi$, one needs to also add more and more *constant* (field-independent additive) counterterms the closer one gets to 8π (one needs to add one more counterterm every time β^2 overcomes a certain threshold, and these thresholds accumulate towards 8π) [34, 35]. A more complete description and a guide to the relevant literature can be found, for example, in section 1 of [36].

Since we work in the region where β is not renormalised, we can fix it to a value of our choice. We will restrict ourselves to $0 < \beta^2 < 8\pi$ for the purposes of this book, as anticipated. This is called the *massive phase* because it has massive soliton and antisoliton excitations. The phase of the theory for $\beta^2 > 8\pi$ is the *massless phase* with massless soliton and antisoliton excitations—as we will shortly see, in this regime the cosine perturbation is irrelevant. For comments on the transition between the two regimes, see for instance [37].

As we have discussed in section 2.2, the theory being integrable allows for the application of exact methods that bypass in principle the perturbative approach and only rely on the structural algebraic properties of the model. Perturbation theory and semiclassical arguments are used as a check of the exact formulas. Moreover, the perturbative expansions also fix some of the parameters that are not constrained when one proceeds purely axiomatically—for instance, the actual renormalised value of the mass can only be ascertained via traditional field-theory methods.

The renormalisation of the Sine-Gordon model is treated in many classic papers, in particular [38–40]. We have mentioned earlier that in theories such as Sine-Gordon, one can get rid of infinities where the theory is renormalisable by imposing the normal-ordering prescription for the Hamiltonian [13, 41]. In Sine-Gordon, this comes at the cost of a multiplicative renormalisation of the infinite bare mass parameter m into a finite parameter m_r appearing in $-\frac{m_r^2}{\beta^2}(:\cos \beta\phi: -1)$, and an infinite shift of the vacuum energy [13, 34, 35]. It is noteworthy that normal-ordering renormalisation still introduces a dependence on a mass scale—a discussion of the significance of the choice of a mass scale in the normal-ordering prescription is contained in [13], see also for instance [42].

In [41], the perturbative contributions to the squared-mass of the elementary Sine-Gordon boson, namely the quantum of the field φ, are analysed. At the leading order in β this would just be m_r^2, and the first correction is found at two loops. We refer to [43] for further discussion on this point.

The exact relationship between the soliton mass m_{sol} (discussed in the next section), which is a quantity of mass-dimension 1, and the coupling m_r was found by Zamolodchikov [44], see also for example [45–47] (see in addition [37]). We express this relationship in the following fashion:

$$-\frac{\mu}{\Lambda_{\mathcal{R}}^{-\frac{\beta^2}{4\pi}}} = \frac{m_0^2}{2\beta^2} = m_{sol}^2 \left(\frac{m_{sol}}{\Lambda_{\mathcal{R}}}\right)^{-\frac{\beta^2}{4\pi}} \frac{\Gamma(\frac{\beta^2}{8\pi})}{\pi\,\Gamma(1-\frac{\beta^2}{8\pi})} \left[\frac{\sqrt{\pi}\Gamma(\frac{1}{2}+\frac{\xi}{2\pi})}{2\Gamma(\frac{\xi}{2\pi})}\right]^{2-\frac{\beta^2}{4\pi}} \to_{\beta\to 0} \frac{m_{sol}^2\beta^2}{128},$$

$$\xi = \frac{\beta^2}{8}\frac{1}{1-\frac{\beta^2}{8\pi}}, \tag{3.3.1}$$

with Λ the mass scale introduced by the renormalisation scheme and μ having mass-dimension $2 - \frac{\beta^2}{4\pi}$. The meaning of the new symbols appearing in (3.3.1) will be explaned shortly: because this is a particularly important point, let us carefully expand it. In this analysis we closely adhere to the original treatment of [44].

To begin with, one should make the following distinction: we can either regard the cosine perturbation as a deformation of the free-boson conformal field theory (CFT), as we have already anticipated, or we can treat the same action in a given perturbative renormalisation scheme—call this scheme $\mathcal{R}$. The result must of course be the same in the two points of view. To make the distinction of the deformation point of view with respect to the perturbative point of view more manifest, we can use the presentation of [44] and explicitly keep track of the mass-dimension carried by the parameters. We introduce the following notation: we use

$$L_{SG} = -\frac{1}{2}\partial_\mu\phi\partial^\mu\phi + 2\mu \cos \beta\phi \tag{3.3.2}$$

if we take the CFT point of view and

$$L_{SG} = -\frac{1}{2}\partial_\mu\phi\partial^\mu\phi + \frac{m_0^2}{\beta^2}[\cos\beta\phi]_{\mathcal{R}} \qquad (3.3.3)$$

if we work in the perturbative renormalisation scheme $\mathcal{R}$. We can explicitly relate the two viewpoints by the formula that we have already displayed above, and which we rewrite here as

$$2\beta^2 \times \frac{\mu}{m_0^2\Lambda_{\mathcal{R}}^{-\frac{\beta^2}{4\pi}}} = 1, \qquad (3.3.4)$$

where $\Lambda_{\mathcal{R}}$ is the mass scale introduced by the renormalisation scheme $\mathcal{R}$ and both sides of (3.3.4) are dimensionless. Once we have defined these scales, and because we can anticipate that there is truly only one physical mass scale in the game—the mass of the quantum soliton m_{sol}—on which all other masses depend, it is therefore expected on dimensional grounds that

$$\mu = \kappa(\beta)\, m_{\text{sol}}^{2-\frac{\beta^2}{4\pi}}, \qquad (3.3.5)$$

for $\kappa(\beta)$ a dimensionless quantity.

The paper [44] provides a very ingenious calculation of the quantity $\kappa(\beta)$. It effectively uses the integral form of the Bethe ansatz for the system coupled to an external gauge field as a clever device to extract the quantity of interest. We shall not reproduce the calculation here because it evades the scopes of this book, but we invite the reader to explore it independently. A very detailed discussion of the calculation is presented in the review [23]. The result that concerns us is the formula

$$\kappa(\beta) = \frac{\Gamma(\frac{\beta^2}{8\pi})}{\pi\,\Gamma(1-\frac{\beta^2}{8\pi})}\left[\frac{\sqrt{\pi}\,\Gamma(\frac{1}{2}+\frac{\xi}{2\pi})}{2\Gamma(\frac{\xi}{2\pi})}\right]^{2-\frac{\beta^2}{4\pi}}, \qquad (3.3.6)$$

which returns (3.3.1). The final step for our line of argument is simply to identify

$$m_r = m_0 \qquad (3.3.7)$$

if we adopt the normal-ordering renormalisation of Coleman's.

From the point of view of perturbation theory, it is natural to see how the exact formula (3.3.1) arises from Feynman graphs. In fact, when combined with formula (3.4.1) below, [44] uses the exact formula to predict the first few orders of the perturbative expansion of the mass of the lightest breather. This should emerge from computing the self-energy correction to the mass of the field quantum in perturbation theory. The result is

$$M_1^2 \sim m_0^2\left(1 + \frac{\beta^2}{8\pi}\left(\gamma + \log\frac{m_0^2}{4\Lambda_{\mathcal{R}}}\right)\right), \qquad (3.3.8)$$

with γ the Euler constant. This is then recognised to be consistent with perturbation theory—undoubtedly upon a choice of the appropriate renormalisation scheme $\mathcal{R}$, see also the comments in [41, 43].

Perturbative and semiclassical evidence for the factorisation of the S-matrix in Sine-Gordon (and in the Thirring model as well) is found in [48–52]. We remind that semiclassical phase-shifts are related to time delays in soliton scattering [53].

Further reading [two days' study]: Become familiar with section 4 in [54].

In the massive phase, the Sine-Gordon theory can be understood as a perturbation of the free-boson CFT by the operator $\cos \beta\phi$ (more references will be given in the Conclusions). The perturbing (composite) operator has a quantum-mechanical scaling dimension $[length]^{-\frac{\beta^2}{4\pi}}$, therefore the coupling in front of it acquires a quantum-mechanical mass-dimension of $2 - \frac{\beta^2}{4\pi}$, as can clearly be seen by (3.3.1). This indeed means that the perturbation is relevant for $0 < \beta^2 < 8\pi$.

This triggers a renormalisation-group flow from the massless UV fixed point (the free boson with central charge $c = 1$) to an infrared (IR) fixed point with $c = 0$ [55]. Such a flow can be described by the TBA, which provides a (particular) c-function $c(L_0) = -\frac{6L_0E_0(L_0)}{\pi}$ (running central charge), where $E_0(L_0)$ is the ground-state energy of the theory on a spatial circle of extent L_0 as computed by the TBA. The calculation will depend on additional scales, such as the soliton mass m_{sol} (see the next section). The parameter L_0 in combination with the masses works as a scale along the flow [56], such that (in unitary theories) the c-function interpolates between c_{UV} and c_{IR} [57], with $c_{UV} > c_{IR}$. For a study away from the massive phase, the reader can consult for instance [55]. In the supplement, as a pedagogical exercise, we will show a simplified example of a TBA which computes the UV and IR central charges in a much simpler model.

The Sine-Gordon model at special values of β reduces to other known integrable models or presents particular extra features:

1. We will have an ample way of motivating that at the value $\beta^2 = 4\pi$ (or in terms of the radius $r_{bos}^2 \equiv \frac{4\pi}{\beta^2} = 1$, see for instance [58–61]) the model develops a free-fermion behaviour.

2. At $\beta^2 = \frac{16\pi}{3}$ (or in terms of the radius $r_{bos}^2 \equiv \frac{4\pi}{\beta^2} = \frac{3}{4}$, in the so-called *repulsive* regime, see the next section) the model develops $\mathcal{N} = 2$ supersymmetry [56, 58, 62–66].

3. As $\beta^2 \to 8\pi^-$, the perturbation tends to the marginal point. In this limit there exists a relationship with the chiral $\mathfrak{su}(2)$ Gross–Neveu model—which is treated in detail for example in [67], see also [60]—and with perturbations of the level-one $\mathfrak{su}(2)$ Wess–Zumino–Witten model. At this point, the symmetry of the system tends to the Yangian symmetry—see for example appendix C of [60].

The reader is also invited to consult [68, 69].

3.3.1 Soliton S-matrix and dressing phase

The main reference for this section is [70].

The spectrum of the quantised Sine-Gordon theory is built out of a soliton and an antisoliton, which may or may not form a bound state. In the classical theory the solitons (kinks) interpolate between the discrete minima of the potential, but their profiles already display particle-like features. This has a precise counterpart in the quantum theory. For the moment let us be content of imagining the spectrum as made up of multi-solitons (and antisolitons) states and their bound states, in addition to the vacuum state $|0\rangle$. We will come back to this point in a later section, when we will discussed the paper by Klassen and Melzer. At that point, it wil be easier to rephrase our considerations in the language of CFT.

In the semiclassical (small β^2) approximation the mass of the soliton and antisoliton is [25, 38–40]

$$m_{\text{sol}} = \frac{m_r}{\xi} = \frac{8m_r}{\beta^2} - \frac{m_r}{\pi}. \tag{3.3.9}$$

The mass (rest-frame energy) was originally calculated in [25]. Formula (3.3.9) consists of the classical contribution plus the first quantum correction from small oscillations, with counterterms suitably subtracted—see also [38–40]. We can see that (3.3.9) gives $m_{\text{sol}} \to \frac{8m_r}{\beta^2}$ at small β^2, which fits with (3.3.1).

The bound state is present in the spectrum when we are in the so-called *attractive regime* $0 < \beta^2 < 4\pi$, while it is absent in the *repulsive* regime $\beta^2 > 4\pi$. The soliton and antisoliton are the 'fundamental' quantum excitations of the model, and correspond to the quantisation of the classical soliton and antisoliton profiles. They are *not* the excitations (quanta) of the field φ itself. In fact, the quantum of the field φ has rather more to do with the soliton-antisoliton bound state, in a way which might become slightly clearer when we shall introduce the duality with the Thirring model, see (5.1.2) and (5.1.3).

The soliton-antisoliton S-matrix is explicitly constructed in [6, 54, 71–73], see [70] for a review. Here we simply report its salient features. In the ordered basis $(s, \bar{s})$, where s stands for the quantum soliton state and $\bar{s}$ for the antisoliton, the S-matrix takes the form

$$S_{SG} = \begin{pmatrix} S & 0 & 0 & 0 \\ 0 & S_T & S_R & 0 \\ 0 & S_R & S_T & 0 \\ 0 & 0 & 0 & S \end{pmatrix}, \tag{3.3.10}$$

where

$$S = S(\theta), \qquad S_T = \frac{\sinh \dfrac{\pi\theta}{\xi}}{\sinh \dfrac{\pi(i\pi - \theta)}{\xi}} S(\theta), \qquad S_R = \frac{i \sin \dfrac{\pi^2}{\xi}}{\sinh \dfrac{\pi(i\pi - \theta)}{\xi}} S(\theta). \tag{3.3.11}$$

The labels T and R stand for *transmission* and *reflection*, respectively, and correspond to the soliton and antisoliton 'flavours' of the scattering particles being either preserved or exchanged. The matrix S_{SG} is written in the basis $\{|s\rangle \otimes |s\rangle, |s\rangle \otimes |\bar{s}\rangle, |\bar{s}\rangle \otimes |s\rangle, |\bar{s}\rangle \otimes |\bar{s}\rangle\}$—it is therefore written as a quantum-group R-matrix (see section 3.6). This means that for instance the S_T entries correspond to $|s(\theta_1)\rangle \otimes |\bar{s}(\theta_2)\rangle \to |s(\theta_1)\rangle \otimes |\bar{s}(\theta_2)\rangle$ and $|\bar{s}(\theta_1)\rangle \otimes |s(\theta_2)\rangle \to |\bar{s}(\theta_1)\rangle \otimes |s(\theta_2)\rangle$. The physicists' way of writing the S-matrix would instead entail that S_T is the amplitude for the processes $|s(\theta_1) \otimes \bar{s}(\theta_2)\rangle^{in} \to |\bar{s}(\theta_2) \otimes s(\theta_1)\rangle^{out}$ and $|\bar{s}(\theta_1) \otimes s(\theta_2)\rangle^{in} \to |s(\theta_2) \otimes \bar{s}(\theta_1)\rangle^{out}$ (making the image of 'transmission' more manifest). In the standard language of quantum-field theoretical S-matrices, we would say for instance that the probability amplitude for a configuration, which started off as $|s(\theta_1)\rangle \otimes |\bar{s}(\theta_2)\rangle^{in}$, to remain in this configuration and therefore simply transmitting through the two particles' 'polarisations' as they were initially, is given by

$$^{out}\langle \bar{s}(\theta_2) \otimes s(\theta_1)|s(\theta_1) \otimes \bar{s}(\theta_2)\rangle^{in} = {}^{out}\langle \bar{s}(\theta_2) \otimes s(\theta_1)|\,\hat{S}\,|\bar{s}(\theta_2) \otimes s(\theta_1)\rangle^{out} = S_T(\theta_1 - \theta_2),$$

having introduced the $\hat{S}$ operator in the standard way (see for instance page 166 of [74])

$$|x\rangle^{in} = \hat{S}\,|x\rangle^{out}. \tag{3.3.12}$$

The topological charge is preserved quantum-mechanically, with the soliton being charged $+1$ and the antisoliton -1. The conservation of the total soliton number is therefore equivalent to the conservation of topological charge and sets to zero any other process (entries of the S-matrix) except those displayed by (3.3.10). In fact the upper $1 \cdot 1$ block of (3.3.10) is the charge $+2$ super-selection sector, the central $2 \cdot 2$ block the charge 0 sector, and the lower $1 \cdot 1$ block the charge -2 sector. Such sectors would be completely independent, if it were not for the crossing-symmetry requirement which mixes them. In addition, the manifest democracy between the soliton and the antisoliton provides an intuitive justification as to why some of the nonzero entries turn out to be equal, while (3.3.11) comes from imposing some of the scattering 'axioms'.

The form (3.3.10) of the S-matrix is traditionally referred to as of *6-vertex* type, because of the number of nonzero entries, by virtue of which it is similar to the 6-vertex model of statistical mechanics [75]. The same will be true for the R-matrix of the XXZ spin-chain to be introduced later on (and it is so for its isotropic limit—the Heisenberg spin-chain—as well).

Notice that the parameter ξ is ill-defined for $\beta^2 = 8\pi$. This is a delicate point and is discussed in [13]. Coleman argued that the energy density of the theory becomes unbounded below if β^2 exceeds 8π. He argued this by means of a perturbative analysis starting from the trivial vacuum $\phi = 0$. It was later remarked [38–40, 54] that the theory can be defined for values of β^2 larger than 8π, provided one perturbatively expands around a nontrivial vacuum [38–40]. Let us once more remark that we shall always assume $\beta^2 < 8\pi$ in this book.

We refer to [70] for the entire step-by-step procedure leading to (3.3.11), which is not yet the complete story. At the end of such a procedure, the overall factor $S(\theta)$

ends up having to solve two conditions dictated by braiding unitarity and crossing symmetry. The two conditions read

$$S(\theta)S(-\theta) = 1, \qquad S(\theta) = \frac{\sinh \dfrac{\pi(i\pi - \theta)}{\xi}}{\sinh \dfrac{\pi\theta}{\xi}} S(i\pi - \theta), \qquad (3.3.13)$$

and they are actually sufficient to determine a *minimal* solution:

$$S(\theta) = -\prod_{k=0}^{\infty} \frac{\Gamma\left(1 + (2k+1)\dfrac{\pi}{\xi} - i\dfrac{\theta}{\xi}\right)\Gamma\left(1 + 2k\dfrac{\pi}{\xi} + i\dfrac{\theta}{\xi}\right)\Gamma\left((2k+1)\dfrac{\pi}{\xi} - i\dfrac{\theta}{\xi}\right)\Gamma\left((2k+2)\dfrac{\pi}{\xi} + i\dfrac{\theta}{\xi}\right)}{\Gamma\left(1 + (2k+1)\dfrac{\pi}{\xi} + i\dfrac{\theta}{\xi}\right)\Gamma\left(1 + 2k\dfrac{\pi}{\xi} - i\dfrac{\theta}{\xi}\right)\Gamma\left((2k+1)\dfrac{\pi}{\xi} + i\dfrac{\theta}{\xi}\right)\Gamma\left((2k+2)\dfrac{\pi}{\xi} - i\dfrac{\theta}{\xi}\right)}. \qquad (3.3.14)$$

The attribute of minimal always refers to having the smallest number of poles and zeroes in the physical strip Im $\theta \in (0, \pi)$.

Exercise [2 days' work]: Prove that (3.3.14) satisfies the second relation in (3.3.13), by using known identities of the Gamma function.

   Guided exercise [2 hours' work]: Reproduce the two different demonstrations given in [70]—pp 33–5.

There exists a well-known integral representation for $S(\theta)$:

$$S(\theta) = -\exp\left[-i \int_0^\infty \frac{dt}{t} \frac{\sinh\left(t(\pi - \xi)\right)}{\sinh(\xi t)\cosh(\pi t)} \sin(2\theta t)\right], \qquad (3.3.15)$$

see also [76]. More precisely we can say that (3.3.14) is the analytic continuation of (3.3.15) to the whole complex θ-plane.

Exercise [1 day's work]: obtain (3.3.15) from (3.3.14) by using the so-called Malmstén integral representation of the Gamma function:

$$\Gamma(x) = \exp \int_0^\infty \frac{dt}{t} \left[(x - 1)e^{-t} + \frac{e^{-tx} - e^{-t}}{1 - e^{-t}}\right], \qquad \text{Re}(x) > 0. \qquad (3.3.16)$$

Identify the region in the θ-plane where (3.3.15) is valid.

   Being a phase factor for real values of θ, as required by physical unitarity, and being an overall factor of the S-matrix, a function such as $S(\theta)$ is often referred to (particularly in the AdS/CFT integrability community) as a *dressing factor*, and its logarithm as a *dressing phase*.

   Notice that the overall minus sign in $S(\theta)$ is at this stage completely arbitrary, since the two conditions (3.3.13) are quadratic in the dressing factor. It is however *very important* that this minus sign is put there by hand, for reasons that will become clear later on. In fact, the condition at equal rapidities $S(0) = -1$, when the two identical scattering particles' 'bare statistics' (i.e. the statistics of the creation operators which creates them from the vacuum) is bosonic, effectively turns them into behaving like fermions. This is what will happen here. As pointed out by Zamolodchikov [14], braiding unitarity for identical particles implies that their scattering amplitude satisfies $S(0)^2 = 1$ hence $S(0) = \pm 1$. This argument applies when the two particles scatter diagonally, i.e., with no transformation of their internal degrees of freedom—otherwise braiding unitarity would involve a sum of terms corresponding to the different channels. When insterted into the Bethe wave-function, this either retains ($S(0) = +1$) or changes ($S(0) = -1$) the bare statistics. An easy way of seeing this is by resorting to a different and simpler model, the nonlinear Schroedinger (or Lieb–Liniger) model—see the supplement. In that case, the two-particle Bethe wave-function can be written as

$$|p_1, p_2\rangle = \int_{-\infty}^{\infty} \int_{-\infty}^{\infty} dx_1 dx_2 \big[\Theta(x_1 - x_2) + S_{LL}(p_1, p_2)\Theta(x_2 - x_1) \big] \times$$
$$\times e^{ip_1 x_1 + ip_2 x_2} \psi^\dagger(x_1)\psi^\dagger(x_2)|0\rangle, \tag{3.3.17}$$
$$S_{LL}(p_1, p_2) = S_{LL}(p_1 - p_2) = \frac{p_1 - p_2 + i\kappa}{p_1 - p_2 - i\kappa},$$

where $p_1 < p_2$ are the real particle momenta (*in* state), $\psi^\dagger$ are the (bosonic) creation operators associated with the perturbative excitations of the field (in that case a complex boson with Galileian invariance), and $|0\rangle$ is actually the perturbative vacuum annihilated by all the ψ's. We have assumed $\kappa > 0$ for simplicity (*repulsive* regime). The convention of denoting by ψ the scalar Lieb–Liniger field is traditional, and we hope that it will not lead to any confusion with the spinorial Thirring field to appear later on. The state (3.3.17) is an exact stationary state—solution of the exact time-independent Schroedinger's equation in the sector where the particle number N, which is conserved, is $N = 2$. It is an *in* state for $p_1 < p_2$ and an *out* state for $p_1 > p_2$, with the *in* and *out* states being rightfully related by the S-matrix S_{LL}—see [77], around formula (2.10). The state (3.3.17) is in principle delta-function normalisable (although we have not bothered reporting the correct normalisation here). The energy eigenvalue is $E = p_1^2 + p_2^2$ (Galileian).

It is now clear to see that, if we set $p_1 = p_2$, the state *vanishes completely*, that is $|p_1, p_1\rangle = 0$ (due to integrating an antisymmetric expression over a symmetric domain, the integrand being antisymmetric as a result of the S-matrix reducing to the value -1 for equal momenta). This works as an effective *exclusion principle*. Notice that the normalisation will not change this outcome[3].

---

[3] To convince oneself that the normalisation does not alter this situation, it is enough to recall that for states in the continuum spectrum we understand the normalisation in the sense of the Dirac delta. It is then simpler to consider the following toy-model wave function (of two genuine free fermions):

$$\frac{1}{2\pi\sqrt{2}}[e^{ip_1 x_1 + ip_2 x_2} - e^{ip_1 x_2 + ip_2 x_1}]. \tag{3.3.18}$$

Such a wave function is delta-function normalised, but it still vanishes for $p_1 = p_2$. In the supplement we shall provide the exact normalisation which makes the eigenstates of the Lieb–Liniger model delta-function normalised, and it will be easy to see that they will not change the conclusions that we have reached here.

To clear any potential surreptitious doubts, it is instructive to rephrase the same exercise in different terms One could approach the two-particle problem for the Lieb–Liniger model by solving an effective Schroedinger problem for a two-particle wave-function, as we will show in the supplement. In terms of this more traditional entirely quantum-mechanical language, the problem consists of solving

$$\left[-\frac{\partial^2}{\partial x_1^2} - \frac{\partial^2}{\partial x_2^2} + 2\kappa\delta(x_1 - x_2) - E \right] f_2(x_1, x_2) = 0. \tag{3.3.19}$$

Because the particles are bosons, we need to impose

$$f_2(x_1, x_2) = f_2(x_2, x_1). \tag{3.3.20}$$

By the familiar trick learnt in the quantum-mechanics class, we assume that the wave function is continuous but not differentiable at the junction $x_1 = x_2$. By integrating the equation (3.3.19) over an infinitesimal segment surrounding $x_1 = x_2$ we obtain the condition

$$\partial_y f(0 +) - \partial_y f(0 -) = \kappa f(0), \tag{3.3.21}$$

where we have found it convenient in this instance to switch to sum and difference (centre of mass and relative coordinate): $x \equiv x_1 + x_2$, $y \equiv x_1 - x_2$, so that the integration is $\int_{0-}^{0+} dy$ and we pick up the discontinuity in the y-derivative in addition to the contribution from the delta potential. Away from the junction clearly $\left[-\frac{\partial^2}{\partial x_1^2} - \frac{\partial^2}{\partial x_2^2} - E \right] f_2(x_1, x_2) = 0$. The solution expressed in the x_1, x_2 variables can be checked to be (without worrying about the normalisation)

$$f_2(x_1, x_2) = e^{ip_1 x_1 + ip_2 x_2} + S_{LL}^{-1}(p_1 - p_2)e^{ip_2 x_1 + ip_1 x_2} \quad \text{if} \quad x_1 < x_2,$$

$$f_2(x_1, x_2) = e^{ip_1 x_2 + ip_2 x_1} + S_{LL}^{-1}(p_1 - p_2)e^{ip_2 x_2 + ip_1 x_1} \quad \text{if} \quad x_2 < x_1,$$

which is symmetric $x_1 \leftrightarrow x_2$ by construction. We can check that (again, modulo normalisation) we consistently have $f(x_1, x_2) \sim \langle 0|\psi(x_1)\psi(x_2)|p_1, p_2\rangle$. The energy E is given by the Galileian expression $p_1^2 + p_2^2$. Now, if we set $p_1 \to p_2$ we obtain

$$\begin{aligned} f_2(x_1, x_2) = 0 \quad &\text{if} \quad x_1 < x_2, \\ f_2(x_1, x_2) = 0 \quad &\text{if} \quad x_2 < x_1, \end{aligned} \tag{3.3.22}$$

which again proves the point about the fermionic behaviour of these interacting bosons—see also the original [78], and again footnote 3.

The Lieb–Liniger model is simpler than Sine-Gordon, in that one can work with the perturbative modes (the Fourier modes of the field) and with relatively little effort construct exact eigenstates of the quantum Hamiltonian using the perturbative modes, by organising them in the fashion of (3.3.17) and its $N > 2$ generalisations. The S-matrix is Galileian-invariant: $S_{LL}(p_1, p_2) = S_{LL}(p_1 - p_2) = \frac{p_1 - p_2 + i\kappa}{p_1 - p_2 - i\kappa}$, for some coupling κ. Clearly $S_{LL}(0) = -1$. The wave-function (3.3.17) vanishes for

$p_1 \to p_2$ precisely because the creation operators $\psi^\dagger$ are bosonic and $S_{LL}(0) = -1$: we have an effective exclusion principle. In the strong κ regime the scattering bosons have an S-matrix that tends to -1 for any value of the momenta: the model has the characteristic behaviour of free fermions (impenetrable bosons or *Tonks-Girardeau* model [79–81]). In that case the wave function reads

$$f_2(x_1, x_2) = e^{ip_1 x_1 + ip_2 x_2} - e^{ip_2 x_1 + ip_1 x_2} \quad \text{if} \quad x_1 < x_2,$$

$$f_2(x_1, x_2) = e^{ip_1 x_2 + ip_2 x_1} - e^{ip_2 x_2 + ip_1 x_1} \quad \text{if} \quad x_2 < x_1.$$

Notice that now $f(x, x) \to 0$ for any x, although we should not forget that the wave function is still *symmetric* (because it is piece-wise defined). Nevertheless, we get the hints of a behaviour typical of fermions, for which the antisymmetry of the wave function would instead be the reason to force it to vanish at equal positions. In addition of course we still have a vanishing wave function at equal momenta. This is sometimes referred to as *fermionisation*.

What we have described is a completely nonperturbative phenomenon: trivially a value of the S-matrix of -1 cannot be expanded as $1 + small$ (perturbation around free particles), as -1 is always a finite distance away from 1. If κ is small, then in perturbation theory we have $S(p_1 - p_2) \sim 1 + 2\sum_{n=1}^{n_0} \left(\frac{i\kappa}{p_1 - p_2}\right)^n$ for some large n_0. We see that we cannot remain perturbative if $p_1 - p_2$ becomes sufficiently small, since the higher order terms in κ start to become comparable in size to the leading order. The exact resummation reveals the particles' fermionic behaviour, since the sum satisfies, as we have seen, $S(0) = -1$. For $\kappa = 0$, i.e. at the leading order in perturbation theory, we would conclude that the particles are bosons—since the sum clearly starts with 1 and the bare statistics is bosonic.

Let us notice that in our integrable world it is not uncommon to encounter *convergent* perturbative series that define analytic functions in the region of convergence, allowing us to then perform analytic continuation. This is a remarkable feature of integrable theories which goes at the heart of their tractability (we thank Zoltan Bajnok for illuminating discussions).

It is interesting to consider that, where we to write (3.3.17) as

$$|p_1, p_2\rangle = \mathcal{A}^\dagger_{p_1} \mathcal{A}^\dagger_{p_2} |0\rangle, \tag{3.3.23}$$

we would have to conclude that the operator $\mathcal{A}^\dagger_p$ satisfies

$$\left(\mathcal{A}^\dagger_p\right)^2 |0\rangle = 0, \tag{3.3.24}$$

which of course has the feeling of creating fermionic-type objects. Such operators can be constructed and lead one naturally into the larger framework of Faddeev-Zamolodchikov operators and the algebraic Bethe ansatz (see later sections). We refer to [77, 82] for the full detail—see for instance formulas (3.81) and (3.83) in [77], taken at $k = k'$.

We summarise our discussion like so: *two bosons (created by operators with bosonic bare statistics, hence commuting operators) equipped with $S(0) = -1$ (when* *assembled into the Bethe states) 'behave' like two fermions (satisfying an exclusion principle).* This is the key observation that allows us to understand the quantisation of the Sine-Gordon (anti)solitons. We refer to reader to [83] for more considerations on this point. Very importantly, 'behaving like' fermions is quite not the same as 'being' fermions, as particular observables (such as certain finite-volume observables for instance [83]) can detect the difference. We will always use the terminology 'behave' in the sense that we have just specified here. For the reader who is more experimentally minded, and is wondering at this point what kind of laboratory experiments (in $1 + 1$ dimensions!) one could ever conceive to tell the difference between bosons which behave like fermions, and fermions which are fermions, we strongly recommend to take a look for example at the ultracold atomic gas measurements reported in [84]. We also suggest [85–88] for example as further reading.

Let us conclude this section by remarking that the infinite product of Gamma functions displayed in the Sine-Gordon dressing factor (3.3.14) converges thanks to the following theorem[4]—see [89] page 178.

Theorem Given N_1 and N_2 complex numbers $\alpha_1, \ldots, \alpha_{N_1}$ and $\beta_1, \ldots, \beta_{N_2}$, respectively, the infinite product

$$\prod_{n=1}^{\infty} \frac{\Gamma(n - \alpha_1)\ldots\Gamma(n - \alpha_{N_1})}{\Gamma(n - \beta_1)\ldots\Gamma(n - \beta_{N_2})} \tag{3.3.25}$$

converges if and only if

1. $N_1 = N_2 = N$ and
2. $\sum_{i=1}^{N} \alpha_i = \sum_{i=1}^{N} \beta_i$ and
3. $\sum_{i=1}^{N} \alpha_i^2 = \sum_{i=1}^{N} \beta_i^2$.

Representing the dressing factor as a convergent infinite product of Gamma functions is ideal to identify its poles and zeros, simply because the Gamma function is never zero and it has poles at negative-integer and zero argument.

Exercise [one afternoon's work]: List all the poles and zeroes of $S(\theta)$, $S_T(\theta)$ and $S_R(\theta)$ in the complex plane, with their respective order.

Notice that remarkably[5] the Sine-Gordon S-matrices (see also the next section) do not carry a dependence on the renormalisation scale, which is confined in the mass parameter and hence in the particles' dispersion relation.

3.4 Breather S-matrix, mixed S-matrix

The poles in the fundamental S-matrix which can give rise to bound states are only those within the physical strip Im $\theta \in (0, \pi)$. In the attractive regime $\xi < \pi$ the bound state spectrum was found by [25] via quantisation of the classical two-soliton solution:

---

[4] We also thank S Frolov for discussions on this point.
[5] We thank Chawakorn Maneerat for pointing this out to us.

$$M_k = 2m_{sol} \sin \frac{k\xi}{2}, \qquad k = 1, 2, \ldots < \frac{8\pi}{\beta^2} - 1 = \frac{\pi}{\xi}. \qquad (3.4.1)$$

Formula (3.4.1) is believed to be exact (see also comments in [54]). As $\beta \to 0$ the breather spectrum becomes continuous, which is the classical result. The small-β expansion of the inverse expression $m_{sol} = \frac{M_1}{2 \sin \frac{\xi}{2}}$ reads

$$m_{sol} \sim \frac{8M_1}{\beta^2} - \frac{M_1}{\pi}, \qquad (3.4.2)$$

which matches the semiclassical result (3.3.9) if we identify at small coupling the lightest breather with the field quantum, in the sense that $M_1 \to m_r$ as $\beta \to 0$.

Guided exercise [4 hours' work]: Reproduce the discussion in [1]—pp 19–21.

Further reading [4 hours' work]: Explore the Coleman–Thun mechanism [1]—pp 21–3 and [90].

As an example, the exact S-matrix of two breathers of mass $M_1 = 2m_{sol} \sin \frac{\xi}{2}$ is given by [41]

$$S_{11} = \frac{\sinh \theta + i \sin \xi}{\sinh \theta - i \sin \xi}. \qquad (3.4.3)$$

This amplitude has simple poles at $\theta = i\xi$, $i(\pi - \xi)$. They correspond to the direct (s-channel) and the crossed (t-channel) appearance of a $k = 2$ breather: for instance, computing the Mandelstam invariant in the direct channel returns

$$(p_1 + p_2)^2 \to 2M_1^2(1 + \cosh \theta_{pole}) = 2M_1^2(1 + \cosh i\xi) = 4M_1^2 \cos^2 \frac{\xi}{2} = 4m_{sol}^2 \sin^2 \xi = M_2^2,$$

where we have used (3.4.1) twice. The simple poles can be easily verified to fall within the physical strip $\mathrm{Im} \ \theta \in (0, \pi)$ as long as ξ is in the attractive regime $\xi \in (0, \pi)$. However, we notice from (3.4.1) that values of ξ slightly lower than π for instance require $k \leqslant 1$, hence the $k = 2$ bound state should not be in the spectrum. In fact this is the case for all $\xi \in (\frac{\pi}{2}, \pi)$. This holds even though we have just found a pole in the physical strip corresponding to this process—although to be completely systematic we should discuss the residue as well, as we have learnt in one of the earlier sections of the book. The point about this precise S-matrix is commented upon in [91], where more in-depth physical considerations are also provided.

It is noteworthy that the amplitude (3.4.3) is a CDD factor[6], see (2.2.3).

Exercise [absorbing as much time as you wish, using hints from [1, 70]]: Have fun doing the same Mandelstam-invariant analysis, as we have just done, for the poles of the various soliton amplitudes.

---

[6] It is also of an identical functional form to the S-matrix of the single bosonic excitation of the so-called *Sinh-Gordon* model, see for instance [92]—we thank Michele Mazzoni for pointing this out to us.

Notice that by being a bound state of a soliton and an antisoliton the breather has zero topological charge and it is a boson, as befits the connection with the fundamental quantum of the scalar field φ. We also remark that at the onset of the repulsive regime all bound states cease to exist, meaning that they are all unstable against the decay into a soliton and an antisoliton. As the coupling increases, and as it is well visible from (3.4.1), the breathers actually disappear one by one from the spectrum (each in turn becoming unstable against decay) every time $\frac{8\pi}{\beta^2} - 1$ passes an integer, until finally only the soliton and the antisoliton are left [25].

There is a certain degree of ambiguity in determining which of the particles is in fact the 'fundamental' quantum—an aspect which has been deemed *nuclear democracy* [93]. This corresponds to the idea that bound states can be conceptually perceived as being as 'fundamental' as their constituents. The paper [25] has a great deal of discussion around the relation between the lowest-lying breather and the quantum of φ. We will see later on that this principle of nuclear democracy applies quite universally. (Anti)solitons are collective excitations of φ, but they can effectively mimic fermionic-like objects and can form bosonic bound states—we could conceive the solitons as 'fundamental' in this respect. On the other hand, the theory describing these bosonic bound states is the theory of the field φ, which perhaps fits even better the bill for a 'fundamental' description. From this perspective, it seems clear that attaching too large a meaning to the word 'fundamental' ends up being rather besides the point.

A more complete account of the S-matrices of Sine-Gordon bound states can be found in [94]—see also [95].

There is a conserved 'parity' charge for a breather of mass M_k, with eigenvalue $(-)^k$, which is related to the transformation $\phi \to -\phi$ under which the classical breathers are 'eigenstates'—see also formulas (4.14a) and (4.14b) in [54], and for example section 5 of [96]. In particular, the lightest breather is odd. See also the discussion in section 4.1 of [24].

3.5 Sine-Gordon and the XXZ spin-chain

There is a very close connection between the Sine-Gordon model and the so-called XXZ spin-chain—a famous one-dimensional model of magnetism. This is the discrete model of spins in the $\frac{1}{2}$ (fundamental) representations of $\mathfrak{su}(2)$, each spin pinned down at one site of a one-dimensional lattice (chain) so as to be each distinguishable from any other, with quantum-mechanical Hamiltonian

$$H_{XXZ} = J\sum_{i=1}^{N}[\sigma_{1,i}\sigma_{1,i+1} + \sigma_{2,i}\sigma_{2,i+1} + \Delta\, \sigma_{3,i}\sigma_{3,i+1}], \tag{3.5.1}$$

where N is the number of lattice sites. We impose periodic boundary conditions—the site $N + 1$ is the site 1. The parameter J is the *exchange coupling*. The name XXZ comes from the fact that the $\sigma_3 - \sigma_3$ coupling is in general different from the other two (the *anisotropy parameter* Δ not being necessarily equal to 1, which would reduce the theory to the Heisenberg model). The chain has nearest-neighbour interactions only—the Hamiltonian density involves only adjacent sites (i and $i + 1$).

This quantum-mechanical spin-chain model is integrable—it directly admits a quantum L-operator from which a quantum monodromy matrix can be built, and N commuting conserved local charges can be obtained from it (in parallel with what happens for classical field theories).

The fact that a quantum-field theory and a spin-chain model are so intimately connected is largely due to the fact that they share a common underlying integrable structure (encoded in a special quantum group which we will be reviewing shortly). It is however still a rather remarkable realisation. To do full justice to such a relationship would require a longer treatment which goes beyond the scope of this book, therefore we will just provide a small guide the literature, while we will focus a bit more on the quantum-group basics in the remainder. For a very nice description with a multitude of detail we refer again to the excellent review [23].

A link between the two models is described in [97]—equations (420)–(449) there. One can engineer a rather involved way of obtaining the Sine-Gordon field-equation from a formal continuous limit of the RTT relations (see later) associated with the XXZ chain, when the L-operator associated with the latter is re-expressed in terms of suitable non-commuting variables. The quantum L-operator can be considered as a quantum analogue of the classical Lax matrix L. The non-commuting variables that one introduces in this context satisfy exchange relations which are reminiscent of those defining the so-called *quantum plane* [98], already anticipating the underlying connection with (the affine version of) $U_q(\mathfrak{su}(2))$—see the next subsections. The XXZ chain therefore can be seen as a lattice regularisation of quantum Sine-Gordon.

In [99]—section 3.1 there—a direct link is drawn between the L-operator of a lattice regularisation of Sine-Gordon and the XXZ L-operator. The two L-operators are essentially seen as arising from different representations of the fundamental quantum group (RTT) relations, in particular they share the same (trigonometric) R-matrix (3.6.14).

A connection has also been established by Destri and de Vega [100, 101] - see also [47] - using the general framework of the six-vertex model. Further analysis of the XXZ spin-chain seen as a lattice regularisation of the Sine-Gordon model can be found in [102–105]. We also refer to [23] for more detail, and the recent [106].

***Project** [6 weeks' work]: Familiarise yourself with [102–105], and in particular study the Bethe equations and the S-matrix analysis. Explore the method of Korepin [107], which is of particular relevance for the Thirring model to be introduced in the next section, and Andrei–Destri [108].*

Finally, we recommend chapter XVIII of [109] for a discussion of the relationship between the XXZ spin-chain and the Sine-Gordon model in the context of bosonisation, and also the recent [110].

3.6 The quantum group $U_q(\mathfrak{su}(2))$

Integrable systems are tightly connected with the mathematical theory of quantum groups, which provides the unified framework to describe them. From the quantum-group viewpoint, there is no difference between a quantum-field theory and a quantum-mechanical spin-chain, as long as their quantum conserved charges define the same *Hopf algebra* (which for our purposes is a synonym of *quantum group*).

For a review of Hopf algebras we refer for example to [98]—see also the recent [111]. The quantum-group symmetry relevant to the Sine-Gordon model is an infinite-dimensional quantum affine algebra, which we will review in the next section. For pedagogical purposes we start with the non-affine version, which allows us to introduce the relevant mathematical concepts.

Here we recall the definition of the Hopf algebra $U_q(\mathfrak{su}(2))$. This Hopf algebra is the deformation of the universal enveloping algebra $U(\mathfrak{su}(2))$ characterised by an (in principle complex) deformation parameter q. It can be presented as the algebra generated by arbitrary polynomials in the three generators E, F, H (positive root generator, negative root generator, and Cartan generator, respectively) subject to the relations ($q \neq \pm 1$)

$$[H, E] = E, \qquad [H, F] = -F, \qquad [E, F] = \frac{q^{2H} - q^{-2H}}{q - q^{-1}}. \qquad (3.6.1)$$

Here we adopt the conventions of [112]. As $q \to 1$, these relations tend to those defining the Lie algebra $\mathfrak{sl}(2)$. One then chooses a suitable real form to obtain $\mathfrak{su}(2)$. The *algebra* structure is given by the multiplication of two generators and by the unit with respect to such multiplication. In the universal envelop we are indeed allowed to write $[a, b] = a . b - b . a$, with $.$ being the multiplication—multiplying generators is in fact the only way we can be allowed to write something like q^H, which is formally a power series. We will often omit to write the sign $.$ for the multiplication. The generators act in various representations, which define the particles of the field theory (or the sites of the spin-chain).

To qualify as a Hopf algebra, we have to exhibit a coproduct for our algebra. This is a map

$$\Delta \colon U_q(\mathfrak{su}(2)) \to U_q(\mathfrak{su}(2)) \otimes U_q(\mathfrak{su}(2)) \qquad (3.6.2)$$

which respects the defining relations of the algebra (algebra homomorphism). The coproduct physically gives the action of the symmetry generators on two particles of a field theory (or two sites of a spin-chain):

$$\Delta(E) = E \otimes q^{-H} + q^H \otimes E, \qquad \Delta(F) = F \otimes q^{-H} + q^H \otimes F,$$
$$\Delta(H) = H \otimes \mathbb{1} + H \otimes \mathbb{1}. \qquad (3.6.3)$$

The identity $\mathbb{1}$ is in a sense the zero-th power of the generators. When $q \to 1$, the coproduct reduces to the standard Leibniz rule for two-particle symmetries. The presence of q consistently *deforms* the standard structure.

In the supplement we explicitly verify that (3.6.3) is an algebra homomorphism—meaning that replacing every generator by its coproduct preserves all the relations (3.6.1). The exponentials of tensor products can be more easily dealt with by using a formal Taylor expansion and noticing that (for bosonic objects) $a \otimes b \cdot c \otimes d = ac \otimes bd$.

The coproduct being an algebra homomorphism implies that, given a representation of $U_q(\mathfrak{su}(2))$, one can generate more representations by repeatedly applying the coproduct. This is nothing else but tensoring representations—it is the same as the usual composition of spins in quantum mechanics, except for the fact that the quantum-mechanical textbook case is always implicitly performed with the standard (Leibniz) coproduct, namely undeformed by q. Whenever we have written

$$\text{total spin} = \text{spin}_1 + \text{spin}_2 \qquad (3.6.4)$$

we have always secretly meant (perhaps without realising it)

$$\Delta(\sigma_i) = \sigma_i \otimes \mathbb{1} + \mathbb{1} \otimes \sigma_i, \qquad (3.6.5)$$

for the ith component of the total spin of two particles.

In the theory of Hopf algebras one also has to define a counit (mapping the Hopf algebra into the field of coefficients, typically $\mathbb{C}$)

$$\epsilon(E) = \epsilon(F) = \epsilon(H) = 0, \qquad \epsilon(\mathbb{1}) = 1, \qquad (3.6.6)$$

which completes the so-called *co-algebra* structure. Finally, the *antipode* map is needed to completely define a Hopf algebra. The antipode map is obtained from the knowledge of the other four maps (multiplication, unit, coproduct and counit) [98]. In the case at hand it reads

$$\Sigma(E) = -q^{-1}E, \qquad \Sigma(F) = -q\,F, \qquad \Sigma(H) = -H, \qquad \Sigma(\mathbb{1}) = \mathbb{1}. \quad (3.6.7)$$

The minus sign is reminiscent of the action of taking the inverse of a Lie-group element (i.e. changing the sign of the corresponding Lie-algebra element); however, the notion is here extended to 'non-Lie' objects. Physically, there is a route from the antipode to the notion of antiparticles [60], which we will have the opportunity to discuss later on.

The $U_q(\mathfrak{su}(2))$ Hopf algebra is *quasi-cocommutative*, meaning that there exists a so-called *R-matrix*. This is an invertible object $R \in U_q(\mathfrak{su}(2)) \otimes U_q(\mathfrak{su}(2))$ satisfying

$$\Delta^{op}(a)R = R\Delta(a), \qquad a = E, F, H, \qquad (3.6.8)$$

where the *opposite* coproduct Δ^{op} is obtained from Δ by permuting the two factors of the tensor product (e.g., $\Delta^{op}(E) = E \otimes q^H + q^{-H} \otimes E$). The relation (3.6.8) says that the coproduct map is not cocommutative (which would be $\Delta^{op} = \Delta$), but it 'almost' is (the Italian *quasi*), because $\Delta^{op}(a) = R\Delta(a)R^{-1}$ for each of the generators $a = E, F, H$.

Physically, we should think of R as a relative of a two-particle S-matrix, precisely $R = perm \circ S$, with *perm* the permutation of the two particles [113]. Having a cocommutative coproduct would reduce (3.6.8) to $[R, \Delta(a)] = 0$, which is the ordinary meaning of a local symmetry of the S-matrix. Formula (3.6.8) generalises this notion to coproducts that are nonlocal, i.e. non-cocommutative.

One of the powerful results in the theory of Hopf algebras is the fact that we can solve (3.6.8) in *any* algebra representation: the standard solution is—see for instance [112]—

$$R = \sum_{n \geq 0} \frac{(q - q^{-1})^n}{[n]_q!} q^{\frac{n(n-1)}{2}} (Fq^H \otimes q^{-H} E)^n q^{2(H \otimes H)}, \tag{3.6.9}$$

where

$$[n]_q! = [n]_q [n-1]_q \ldots [2]_q [1]_q, \qquad [n]_q = \frac{q^n - q^{-n}}{q - q^{-1}}. \tag{3.6.10}$$

This formula is called the *universal R-matrix* precisely because it does not care about the representation and it works purely based on the commutation relations. It also satisfies the Yang–Baxter equation purely based on the commutation relations.

Exercise [to do once in a lifetime]: Verify (3.6.8) using only the formula for the universal R-matrix (3.6.9)–(3.6.10) and the defining relations of $U_q(\mathfrak{su}(2))$.

Once the cocommutativity relation has been established for the three generators, it will follow for any polynomial thereof simply by homomorphism. The striking feature is that we can now evaluate (3.6.9) in any of the representations of $U_q(\mathfrak{su}(2))$—each time obtaining a different matrix solution of the Yang–Baxter equation. Each of these different matrices defines a different integrable scattering problem, but all of these problems are still connected with the same underlying algebraic structure. They can also define different integrable models altogether, and those integrable models will still be connected by their algebraic binding. For example, by plugging the fundamental representation

$$E = \sigma_+ = \frac{1}{2}(\sigma_1 + i\sigma_2), \qquad F = \sigma_- = \frac{1}{2}(\sigma_1 - i\sigma_2), \qquad H = \frac{1}{2}\sigma_3, \tag{3.6.11}$$

into (3.6.9)–(3.6.10), one obtains the R-matrix

$$R = \begin{pmatrix} \sqrt{q} & 0 & 0 & 0 \\ 0 & \frac{1}{\sqrt{q}} & 0 & 0 \\ 0 & \frac{q - \frac{1}{q}}{\sqrt{q}} & \frac{1}{\sqrt{q}} & 0 \\ 0 & 0 & 0 & \sqrt{q} \end{pmatrix}. \tag{3.6.12}$$

Notice that the representation (3.6.11) does not depend on q, and it is therefore also a representation of $\mathfrak{sl}(2)$. This is accidental and it is a peculiarity of very special representations such as the fundamental representation.

In the supplement, exploiting the fact that in the fundamental representation E and F are nilpotent matrices, we show how to obtain (3.6.12) from (3.6.9).

Exercise [30 minutes' work using computer algebra]: Check that (3.6.8) is satisfied in the fundamental representation.

Most of the 'axioms' of the S-matrix programme can be incorporated into the framework of Hopf algebras, and in this way they become amenable to an analysis based on methods of representation theory. In fact, the theory of quantum groups was originally developed by Drinfeld as a way of *generating* solutions of the Yang–Baxter equation via an algebraic machinery. Such solutions are ready to then become the S-matrices of integrable systems because they naturally satisfy the 'axioms'—in fact all such S-matrices descend from the universal R-matrix which satisfies the following properties (guaranteed by a number of theorems—see [98] for the hypotheses and the proofs):

$$RR^{op} = 1 \otimes 1, \qquad (\Sigma \otimes 1)R = R^{-1} = (1 \otimes \Sigma^{-1})R,$$
$$R_{12}R_{13}R_{23} = R_{23}R_{13}R_{12}, \qquad (\Delta \otimes 1)R = R_{13}R_{23}, \qquad (1 \otimes \Delta)R = R_{13}R_{12}. \tag{3.6.13}$$

By representing the relations (3.6.13) in specific representations one recovers the S-matrix constraints (resp., braiding unitarity, crossing, Yang–Baxter equation and bootstrap), which we have described at the beginning of this book [113]. In particular, the antipode has something to do with antiparticle representations and the coproduct featuring in (3.6.13), as we have motivated earlier, has something to do with the tensor product of two representations (which physically allows us to extract the bound state as a particular irreducible component).

It is important to mention that the universal R-matrix has been systematically constructed for all the trigonometric and rational quantum groups based on simple Lie (super)algebras, thanks primarily to a series of work by the mathematicians S. Khoroshkin and V. Tolstoy.

Versions of the R-matrix (3.6.12) appear in the study of the RTT relations of the XXZ spin-chain (see for instance section 10 in [97]). Reference [97] provides a very instructive description of the RTT relations and the method of the algebraic Bethe ansatz for spin-chains. The introduction of nonconstant parameters (inhomogeneities on spin-chain sites) can typically be performed via the process of *Baxterisation*, or by switching to the affine version of the quantum group $U_q(\hat{\mathfrak{sl}}(2))$ (see the next section). For example, one can prove that the R-matrix

$$R(\theta) = \frac{1}{\sinh(\eta + \theta)} \begin{pmatrix} \sinh(\eta + \theta) & 0 & 0 & 0 \\ 0 & \sinh\theta & e^{\theta}\sinh\eta & 0 \\ 0 & e^{-\theta}\sinh\eta & \sinh\theta & 0 \\ 0 & 0 & 0 & \sinh(\eta + \theta) \end{pmatrix} \tag{3.6.14}$$

(in a convenient normalisation), dependent on the additional Baxterised variable θ for any fixed $\eta \in \mathbb{C}$, satisfies (3.6.8) with respect to a version of the $U_q(\mathfrak{sl}(2))$ quantum-group symmetry given by

$$\Delta(E) = E \otimes \mathbb{1} + q^{-2H} \otimes E, \qquad \Delta(F) = F \otimes q^{2H} + \mathbb{1} \otimes F,$$
$$\Delta(H) = H \otimes \mathbb{1} + \mathbb{1} \otimes H, \tag{3.6.15}$$

in the representation (3.6.11), and having set

$$q = e^{\eta}. \tag{3.6.16}$$

The parameter q ultimately depends on the anisotropy parameter Δ in the XXZ Hamiltonian, but not on the exchange coupling J. One can find in the literature that the anisotropy parameter is connected with $\frac{q+q^{-1}}{2}$. As $q \to 1$ one obtains the Heisenberg spin-chain, displaying an undeformed $\mathfrak{su}(2)$ symmetry (which also extends to the Baxterised version, namely the Yangian of $\mathfrak{su}(2)$ [67]). In the Heisenberg spin-chain, the sign of J determines whether the ground state is ferromagnetic, namely with all the spins aligned, or antiferromagnetic, namely having a maximum disalignment.

Exercise [3 hours' work]: Verify that also (3.6.15) is a legitimate coproduct for $U_q(\mathfrak{sl}(2))$, in particular that it respects the homomorphism property. Then verify that (3.6.14) is an R-matrix for this coproduct (i.e., it is such that the cocommutativity property is verified).

Exercise [1 hours' work]: Write a symbolic computer programme which verifies that (3.6.14) satisfies the Yang–Baxter equation. For this it will be useful to write $\theta = \theta_1 - \theta_2$, such that R_{12} will be a function of $\theta_1 - \theta_2$, while R_{13} will be a function of $\theta_1 - \theta_3$, etc.

3.6.1 Towards quantum affine

The R-matrix (3.6.14) directly controls the RTT relations of the XXZ spin-chain, see for instance [114]—formula (4.11) there. This R-matrix is of the same type as the R-matrix of the Sine-Gordon soliton–antisoliton scattering (see also chapter 18 in [28]), both belonging to the family of trigonometric quantum-group R-matrices [27]. The association of the Sine-Gordon model with the quantum group $U_q(\widehat{\mathfrak{sl}(2)})$ is well known and can in fact be purely derived via field-theoretic methods—see [60] (particularly their formulas (3.39a)–(3.39c) and their appendix B) and [115]. One can explicitly verify that the Sine-Gordon S-matrix S_{SG}—in the form (3.3.10), (3.3.11), which we remind is written as a quantum-group R-matrix—satisfies for example

$$\Delta^{op}(\hat{E}) \, S_{SG} = S_{SG} \, \Delta(\hat{E}), \tag{3.6.17}$$

with

$$\Delta(\hat{E}) = \hat{E} \otimes \mathbb{1} + q^{-2H} \otimes \hat{E}, \tag{3.6.18}$$

and $\hat{E}$ in the representation pertaining to the particle i, with $i = 1, 2$, being given by $x_i E_{12}$, having set $x_i = \exp[\frac{\pi\theta_i}{\xi}]$. The parameter q is related[7] to the dimensionless coupling β by

$$q = \exp\left[\frac{8\pi^2 i}{\beta^2}\right]. \tag{3.6.19}$$

Notice that the 'free-fermion' point $\beta^2 = 4\pi$, namely $q = 1$, corresponds to an undeformed coproduct satisfying $\Delta^{op}(\hat{E}) = \Delta(\hat{E})$ (cocommutative). We remark that $\hat{E}$ is formally constructed as a particular combination of the standard basis generators of $U_q(\hat{sl}(2))$—see for instance formula (3.22) in [60].

At the special supersymmetric value $\beta^2 = \frac{16\pi}{3}$ we see that $q = -i$. We now need to recall the eigenvalues of H in the fundamental representation are $\pm\frac{1}{2}$, to conclude that the coproduct develops an interesting effect. Specifically, we can see that for example

$$\Delta(\hat{E})|1\rangle \otimes |2\rangle = (\hat{E} \otimes \mathbb{1} + q^{-2H} \otimes \hat{E})|1\rangle \otimes |2\rangle = (-1)^{\frac{1}{2}} x_2 |1\rangle \otimes |1\rangle + 0, \tag{3.6.20}$$

where we have used the fact that $\hat{E} = x_i E_{12}$ on the ith particle in the fundamental representation, $i = 1, 2$, and that $q = -i$ at the special point which we are studying. Both $|1\rangle$ and $|2\rangle$ have the same fermionic number $|1| = |2| = 0$, and therefore $\hat{E}$ has zero fermionic number—because it sends the state $|2\rangle$ into the state $|1\rangle$. Notice that, as we have discussed earlier in this section, the states $|1\rangle = |s\rangle$ and $|2\rangle = |\bar{s}\rangle$ are bosons with S-matrix $S(0) = -1$, hence they behave like fermions but their bare statistics (which are the statistics of the vectors $|s\rangle$ and $|\bar{s}\rangle$) are zero.

We can now see that (3.6.20) is the same as if we had

$$\Delta(\hat{E}')|1'\rangle \otimes |2'\rangle = (\mathbb{1} \otimes \hat{E}' + \hat{E}' \otimes \mathbb{1})|1'\rangle \otimes |2'\rangle = (-1)^{\frac{1}{2}} x_2 |1'\rangle \otimes |1'\rangle + 0 \tag{3.6.21}$$

The identification of the outcomes in (3.6.20) and (3.6.21) as being the same numerical vector comes about if make the following assignment: first, $\hat{E}'$ is still represented as $E_{12} = \begin{pmatrix} 0 & 1 \\ 0 & 0 \end{pmatrix}$, but has fermionic number $|\hat{E}'| = 1$; second, $|1'\rangle$ is still represented as $\begin{pmatrix} 1 \\ 0 \end{pmatrix}$, but has fermionic number $|1'| = \frac{1}{2}$; third, $|2'\rangle$ is still represented as $\begin{pmatrix} 0 \\ 1 \end{pmatrix}$, but has fermionic number $|2'| = -\frac{1}{2}$. In this way, we can write (3.6.21) as

$$\Delta(\hat{E}')|1'\rangle \otimes |2'\rangle = (-1)^{\frac{1}{2}}|1'\rangle \otimes |1'\rangle = (-)^{|\hat{E}'||1'|}|1'\rangle \otimes \hat{E}'|2'\rangle, \tag{3.6.22}$$

---

[7] Notice that this connection bears a certain degree of ambiguity, and different sources present different deformation parameters q with slight variations in the coupling-constant dependence. It is, however, clear that a relationship of the type (3.6.19) will hold.

which is exactly the way one proceeds throughout (3.6.21), and which numerically coincides with (3.6.20).

Likewise, we can see that

$$\Delta(\hat{E})|2\rangle \otimes |2\rangle = (\hat{E} \otimes 1 + q^{-2H} \otimes \hat{E})|2\rangle \otimes |2\rangle = (-1)^{-\frac{1}{2}} x_2 \, |2\rangle \otimes |1\rangle + 0, \quad (3.6.23)$$

which can be equivalently seen as emerging from swapping an operator with a fermionic number 1 and a state with fermionic number $-\frac{1}{2}$.

This exercise is teaching us something very interesting. On the one hand, we have rephrased the action of the coproduct of the quantum-group generators into an equivalent form, which involves acting with a new generator that changes the fermionic number of the states by 1. This new generator is therefore a *supercharge* in supersymmetry parlance. In ordinary supersymmetry a supercharge changes a boson into a fermion, and vice versa, while here it adds 1 to the fermionic number of a state with fermionic number $-\frac{1}{2}$ to obtain a state with fermionic number $\frac{1}{2}$. These vectors are now interpreted as providing a new way of describing the same numerical result that we had before, but in terms of states with a fractional fermionic number. On the other hand, we see that the action of the symmetry has changed from being nonlocal—as in (3.6.20) where there are nontrivial generators acting on both states at the same time—to being local—as in (3.6.21) where we always have the identity on at least one of the states, signalling that we are acting locally on one particle at a time.

This is telling us that the model has developed supersymmetry (which turns out to be extended to $\mathcal{N} = 2$) at this special value of the coupling. The reader is also referred to [116], where this is then generalised to an entire family of special values of q. It is, however, important to keep in mind the fashion in which the supersymmetry has come to be. This is discussed with great detail in [56, 58, 62–66], in particular in section 4.1 of [64]. The description in which the supersymmetry is locally realised, and which is a physical realisation of the situation (3.6.21), is provided by the particular $\mathcal{N} = 2$ theory, which is detailed in [64] in the shape of an effective Landau–Ginzburg model. The states there are the solitons of that particular model. In the Sine-Gordon theory, at this special value of the coupling the same numerical effect is incarnated by (3.6.20), which, although giving the same result, sees a bosonic symmetry being nonlocally realised. The local realisation of the soliton/antisoliton fermion number in Sine-Gordon is still provided by the topological charge, which is additive (i.e., it acts via a local coproduct). Despite the coincident numerical outcomes, this still warns us that the particles in the two models (the $\mathcal{N} = 2$ theory theory of [64] and the Sine-Gordon model at $\beta^2 = \frac{16\pi}{3}$) are not exactly the same particles—they just behave in the same way.

Notice that another symmetry of the Sine-Gordon S-matrix can be checked to be

$$\Delta(\hat{F}) = \hat{F} \otimes 1 + q^{2H} \otimes \hat{F}, \quad (3.6.24)$$

with $\hat{F}$ being given by $x_i E_{21}$. For this symmetry completely analogous considerations apply as we have displayed above, this time with $|\hat{F}'| = -1$.

3.6.2 Project

Project [5 weeks' work]: Explore the literature of fractional fermion numbers in 1 + 1 dimensions, taking the moves from [56, 58, 62–66] but also from [117]. Revisit and expand the considerations we have made in this section in the light of such a literature review.

3.7 The quantum affine symmetry

We have spoken of the process of Baxterisation as a way of introducing a dependence on the spectral parameter into an R-matrix. A more natural process from the point of view of quantum-group theory is to consider infinite-dimensional quantum groups, such as the Yangian—which we will discuss in the supplement. Another type of infinite-dimensional quantum groups is provided by quantum affine algebras. The latter are relevant for Sine-Gordon, in particular the affine version of $U_q(\mathfrak{sl}(2))$, which is often denoted as $U_q(\hat{\mathfrak{sl}}(2))$. In infinite-dimensional quantum groups, what typically happens is that the spectral parameter provides an extra direction, an 'affine' direction, which produces an infinite tower of levels by Taylor (Laurent) expansion.

In the supplement we will explain, using the example of the Yangian, the difference between the different traditional realisations of infinite-dimensional quantum groups. Here we focus on the so-called 'second realisation' of $U_q(\hat{\mathfrak{sl}}(2))$—see for instance [118, 119]. This realisation epresses the quantum group in terms of the following generators:

$$\left\{ d, c, k, k^{-1}, e_n, f_n \right\} \tag{3.7.1}$$

(where effectively one can write $k^{\pm 1} = q^{\pm h}$), organised in the generating functions

$$e(z) = \sum_{n=-\infty}^{\infty} e_n z^n, \qquad f(z) = \sum_{n=-\infty}^{\infty} f_n z^n,$$

$$\psi^{\pm} = \sum_{n=0}^{\infty} \psi_n^{\pm} z^{\mp n} = k^{\pm 1} \exp \left[\pm (q - q^{-1}) \sum_{n>0} a_{\pm n} z^{\mp n} \right], \tag{3.7.2}$$

and subject to a number of defining relations. Such relations are listed in equations (2.6a)–(2.6k) of [118, 119], where one can also find the coproduct—equations (2.8a)–(2.8c). It is quite rare to have explicit formulas for the coproduct of an infinite-dimensional quantum group, typically because the complication of the 'tail' (see the supplement) grows with the level. In the case of infinite-dimensional quantum groups where the level $n = 0$ is either $\mathfrak{sl}(2)$ or $U_q(\mathfrak{sl}(2))$, one sometimes finds explicit expressions organised in terms of generating functions: it is still necessary to expand the coproducts of the generating functions to obtain the ever more complicated coproduct of the modes, but at least one has at one's disposal a set of exact formulas.

The universal R-matrix is also explicitly reported in [118, 119]. It is factorised according to the familiar root decomposition and assembled in terms of known factors. These factors are also explicitly reported in [120]. The idea is now that the fundamental representation can be extended to the affine algebra, and inserted into the formula for the universal R-matrix. The result gives an alternative access to the (Baxterised) R-matrix that we have discussed in the context of the XXZ spin-chain. In fact, it is sufficient to take the formula reported in [120], equation (8)—ignoring the dressing factor f_q (i.e., with f_q set to 1)—and set the variables according to the following assignment:

$$x_{GZ} = e^{-2\theta_1}, \qquad y_{GZ} = e^{-2\theta_2}, \qquad q_{GZ} = e^{\eta}, \tag{3.7.3}$$

where the suffix $_{GZ}$ stands for the variables used in [120] (Gould and Zhang), to obtain perfect agreement with our equation (3.6.14).

Exercise [0.5 hours' work]: Verify the statement that has just been made. You should bear in mind that equation (8) in [120] is written is four-dimensional notation; therefore, Gould and Zhang denote by e_{11} the matrix that sends the first four-dimensional basis state into itself—we call this state $|s\rangle \otimes |s\rangle$ or $|1\rangle \otimes |1\rangle$ (setting $|1\rangle \equiv |s\rangle$ and $|2\rangle \equiv |\bar{s}\rangle$), so that the conversion goes $e_{11} = E_{11} \otimes E_{11}$. Likewise, Gould and Zhang's e_{22} is our $E_{11} \otimes E_{22}$, e_{33} is our $E_{22} \otimes E_{11}$, e_{44} is our $E_{22} \otimes E_{22}$, e_{32} is our $E_{21} \otimes E_{12}$, and finally e_{23} is our $E_{12} \otimes E_{21}$.

The XXZ spin-chain tends to the Heisenberg spin-chain in the *isotropic* limit; namely, when the anistropy parameter Δ tends to 1. In this limit the symmetry algebra also tends to an *isotropic* (or *rational*) version, which is the Yangian. As we have recalled, we will describe the Yangian in great detail in the supplement. Let us also point out that an alternative presentation of the quantum affine algebra $U_q(\widehat{\mathfrak{sl}}(2))$ can be contained, for instance, in [116].

References

[1] Dorey P 1996 *Eotvos Summer School in Physics: Conformal Field Theories and Integrable Models* (Berlin: Springer) **8** 85–125 pp

[2] Ablowitz M J, Kaup D J, Newell A C and Segur H 1973 Nonlinear-evolution equations of physical significance *Phys. Rev. Lett.* **31** 125–7

[3] Faddeev L D and Takhtajan L A 1974 Essentially non-linear one-dimensional model of the classical field theory *Teor. Mat. Fiz.* **21** 160–74

[4] Mclaughlin D W 1975 Four examples of the inverse method as a canonical transformation *J. Math. Phys.* **16** 96–9

[5] Flume R 1976 Anomalies of currents in the quantized sine-Gordon equation *Phys. Lett.* B **62** 93

[6] Kulish P P and Nissimov E R 1976 Conservation laws in the quantum theory: cos ϕ in two-dimensions and in the massive Thirring model *JETP Lett.* **24** 220–3

[7] Luscher M 1976 Dynamical charges in the quantized, renormalized massive Thirring model *Nucl. Phys.* B **117** 475–92

[8] Flume R, Mitter P K and Papanicolaou N 1976 Higher conserved charges in the massive Thirring model *Phys. Lett.* B **64** 289–92

[9] Flume R and Meyer S 1977 Renormalization of a higher conservation law in the massive Thirring model *Lett. Nuovo Cim.* **18** 238

[10] Luscher M 1978 Quantum nonlocal charges and absence of particle production in the two-dimensional nonlinear sigma model *Nucl. Phys.* **B 135** 1–19

[11] Nissimov E R 1977 Infinite set of conservation laws in the quantum sine-Gordon and the massive thirring models *Bulg. J. Phys.* **4** 113–24

[12] Luscher M and Pohlmeyer K 1978 Scattering of massless lumps and nonlocal charges in the two-dimensional classical nonlinear sigma model *Nucl. Phys.* **B 137** 46–54

[13] Coleman S R 1975 The quantum sine-Gordon equation as the massive thirring model *Phys. Rev.* D **11** 2088

[14] Zamolodchikov A B 1990 Thermodynamic Bethe ansatz in relativistic models. Scaling three state Potts and Lee-Yang models *Nucl. Phys.* **B 342** 695–720

[15] Fioravanti D, Mariottini A, Quattrini E and Ravanini F 1997 Excited state Destri-De Vega equation for sine-Gordon and restricted sine-Gordon models *Phys. Lett.* **B 390** 243–51

[16] Feverati G, Ravanini F and Takacs G 1998 Scaling functions in the odd charge sector of sine-Gordon/massive Thirring theory *Phys. Lett.* **B 444** 442–50

[17] Feverati G, Ravanini F and Takacs G 1999 Nonlinear integral equation and finite volume spectrum of sine-Gordon theory *Nucl. Phys.* **B 540** 543–86

[18] Destri C and De Vega H J 1997 Non-linear integral equation and excited-states scaling functions in the sine-Gordon model *Nucl. Phys.* **B 504** 621–64

[19] Al B 1991 Zamolodchikov. On the thermodynamic Bethe ansatz equations for reflectionless ade scattering theories *Phys. Lett.* **B 253** 391–4

[20] Tateo R 1995 New functional dilogarithm identities and sine-Gordon Y-systems *Phys. Lett.* **B 355** 157–64

[21] Nagy B C, Kormos M and Takács G 2023 Thermodynamics and fractal Drude weights in the sine-Gordon mode *Phys. Rev.* **B 108** L241105

[22] Nagy B C, Takács G and Kormos M 2023 *Thermodynamic Bethe ansatz and generalised hydrodynamics in the sine-Gordon model* arXiv:2312.03909v3 [cond-mat.str-el]

[23] Bajnok Z and Samaj L 2011 Introduction to integrable many-body systems III *Acta Phys. Slovaca* **61** 129–271

[24] Klassen T R and Melzer E 1993 Sine-Gordon not equal to massive Thirring, and related heresies *Int. J. Mod. Phys.* A **8** 4131–74

[25] Dashen R F, Hasslacher B and Neveu A 1975 The particle spectrum in model field theories from semiclassical functional integral techniques *Phys. Rev.* D **11** 3424

[26] Manton N and Stuart D Classical and Quantum Solitons Based on Cambridge Lectures, notes taken by D. Chua

[27] Torrielli A 2016 Lectures on classical integrability *J. Phys.* A **49** 323001

[28] Mussardo G 2020 *Statistical Field Theory. Oxford Graduate Texts* (Oxford: Oxford University Press) 3 p

[29] Ruijsenaars S N M and Schneider H 1986 A new class of integrable systems and its relation to solitons *Ann. Phys.* **170** 370–405

[30] Ruijsenaars S N M 2001 Sine-Gordon solitons versus relativistic Calogero–Moser particles *Integrable Structures of Exactly Solvable Two-Dimensional Models of Quantum Field Theory* (Dordrecht: Springer) 273–92 pp

[31] Amit D J, Goldschmidt Y Y and Grinstein S 1980 Renormalisation group analysis of the phase transition in the 2D Coulomb gas, sine-Gordon theory and xy-model *J. Phys. A: Math. Gen.* **13** 585

[32] Tong D *Cambridge Part III Lecture Notes*

[33] Gogolin A O, Nersesyan A A and Tsvelik A M 2004 *Bosonization and Strongly Correlated Systems* (Cambridge: Cambridge University Press)

[34] Benfatto G, Gallavotti G and Nicolo F 1982 On the massive sine-Gordon equation in the first few regions of collapse *Commun. Math. Phys.* **83** 387–410

[35] Nicolò F 1983 On the massive sine-Gordon equation in the higher regions of collapse *Commun. Math. Phys.* **88** 581–600

[36] Fröb M B and Cadamuro D 2022 Local operators in the sine-Gordon model: $\partial_\mu \phi \, \partial_\nu \phi$ and the stress tensor arXiv preprint arXiv:2205.09223

[37] Daviet R and Dupuis N 2019 Nonperturbative functional renormalization-group approach to the sine-Gordon model and the Lukyanov-Zamolodchikov conjecture *Phys. Rev. Lett.* **122** 155301

[38] Faber M and Ivanov A N 2001 On the equivalence between sine-Gordon model and Thirring model in the chirally broken phase of the Thirring model *Eur. Phys. J.* C **20** 723–57

[39] Faber M and Ivanov A N 2003 Is the energy density of the ground state of the sine-Gordon model unbounded from below for beta**2 greater than 8 pi? *J. Phys. A* **36** 7839

[40] Bozkaya H, Faber M, Ivanov A N and Pitschmann M 2006 On the renormalization of the two-point Green function in the sine-Gordon model *J. Phys. A* **39** 2177–201

[41] Karowski M and Weisz P 1978 Exact form-factors in (1 + 1)-dimensional field theoretic models with soliton behavior *Nucl. Phys.* B **139** 455–76

[42] Rychkov S and Vitale L G 2016 Hamiltonian truncation study of the ϕ^4 theory in two dimensions. II. The Z_2-broken phase and the Chang duality *Phys. Rev. D* **93** 065014

[43] Babujian H and Karowski M 1999 The exact quantum sine-Gordon field equation and other non-perturbative results *Phys. Lett.* B **471** 53–7

[44] Zamolodchikov A B 1995 Mass scale in the sine-Gordon model and its reductions *Int. J. Mod. Phys. A* **10** 1125–50

[45] Konik R M and LeClair A 1996 Short distance expansions of correlation functions in the sine-Gordon theory *Nucl. Phys.* B **479** 619–53

[46] Lukyanov S L and Zamolodchikov A B 1997 Exact expectation values of local fields in quantum sine-Gordon model *Nucl. Phys.* B **493** 571–87

[47] Feverati G 2000 Finite volume spectrum of sine-Gordon model and its restrictions **1** University of Bologna

[48] Arefeva I and Korepin V E 1974 Scattering in two-dimensional model with Lagrangian (1/gamma) ((d(mu)u)**2/2 + m**2 cos(u-1)) *Pisma Zh. Eksp. Teor. Fiz* **20** 680

[49] Korepin V E and Faddeev L D 1975 Quantization of solitons *Theor. Math. Phys.* **25** 1039–49

[50] Berg B, Karowski M and Thun H J 1976 Conserved currents in the massive Thirring model *Phys. Lett.* B **64** 286–8

[51] Berg B, Karowski M and Thun H J 1977 A higher conserved current in the quantized massive Thirring model *Nuovo Cim. A* **38** 11

[52] Berg B 1977 The massive Thirring model: particle scattering in perturbation theory *Nuovo Cim. A* **41** 58

[53] Jackiw R and Woo G 1975 Semiclassical scattering of quantized nonlinear waves *Phys. Rev. D* **12** 1643

[54] Zamolodchikov A B and Zamolodchikov A B 1979 Factorized S matrices in two-dimensions as the exact solutions of certain relativistic quantum field models *Ann. Phys.* **120** 253–91

[55] Bacsó V, Defenu N, Trombettoni A and Nándori I 2015 c-Function and central charge of the sine-Gordon model from the non-perturbative renormalization group flow *Nucl. Phys.* B **901** 444–60

[56] Fendley P and Saleur H 1992 N = 2 Supersymmetry, Painleve III and exact scaling functions in 2-D polymers *Nucl. Phys.* B **388** 609–26

[57] Fendley P, Saleur H and Zamolodchikov A B 1993 Massless flows. 1. The sine-Gordon and O(n) models *Int. J. Mod. Phys.* A **8** 5717–50

[58] Dunning T C 2000 Perturbed conformal field theory, nonlinear integral equations and spectral problems *PhD thesis* Durham University

[59] Ravanini F 2001 Finite size effects in integrable quantum field theories *In Non-perturbative QFT Methods and Their Applications* (Singapore: World Scientific) 199–264 pp

[60] Bernard D and Leclair A 1991 Quantum group symmetries and nonlocal currents in 2-D QFT *Commun. Math. Phys.* **142** 99–138

[61] Tong D *Quantum Field Theory on a Line* https://www.damtp.cam.ac.uk/user/tong/gauge-theory/72d.pdf (accessed March 25 2024)

[62] Ahn C, Bernard D and LeClair A 1990 Fractional supersymmetries in perturbed coset Cfts and integrable soliton theory *Nucl. Phys.* B **346** 409–39

[63] Leclair A 1991 Infinite quantum group symmetry in 2-d quantum field theory *International Conference on Differential Geometric Methods in Theoretical Physics* Vol. **1**

[64] Fendley P and Intriligator K A 1992 Scattering and thermodynamics in integrable N = 2 theories *Nucl. Phys.* B **380** 265–90

[65] Fendley P and Intriligator K A 1994 Exact N = 2 Landau-Ginzburg flows *Nucl. Phys.* B **413** 653–74

[66] Fendley P 1998 Excited state energies and supersymmetric indices *Adv. Theor. Math. Phys.* **1** 210–36

[67] Loebbert F 2016 Lectures on Yangian symmetry *J. Phys.* A **49** 323002

[68] Freedman D Z and Pilch K 1988 Thirring model partition functions and harmonic differentials *Phys. Lett.* B **213** 331–6

[69] Freedman D Z and Pilch K 1989 Thirring model on a Riemann surface *Ann. Phys.* **192** 331

[70] Bombardelli D 2016 S-matrices and integrability *J. Phys.* A **49** 323003

[71] Faddeev L D and Takhtajan L A 1974 Essentially nonlinear one-dimensional model of the classical field theory *Technical report, Joint Inst. for Nuclear Research*

[72] Zamolodchikov A B 1976 Exact s-matrix of quantum sine-Gordon solitons. Technical report, Gosudarstvennyj Komitet po Ispol'zovaniyu Atomnoj Ehnergii SSSR

[73] Zamolodchikov A B 1977 Exact two-particle s-matrix of quantum sine-Gordon solitons *Commun. Math. Phys.* **55** 183–6

[74] Itzykson C and Zuber J-B 2012 *Quantum Field Theory* Courier Corporation

[75] Reshetikhin N 2010 Lectures on the integrability of the six-vertex model *Exact Methods in Low-dimensional Statistical Physics and Quantum Computing* (Oxford: Oxford Univ. Press) 197–266 pp

[76] Karowski M, Thun H J, Truong T T and Weisz P H 1977 On the uniqueness of a purely elastic s matrix in (1+1)-dimensions *Phys. Lett.* B **67** 321–2

[77] Thacker H B 1981 Exact integrability in quantum field theory and statistical systems *Rev. Mod. Phys.* **53** 253

[78] Lieb E H and Liniger W 1963 Exact analysis of an interacting bose gas. i. the general solution and the ground state *Phys. Rev.* **1963** 1605

[79] Tonks L 1936 The complete equation of state of one, two and three-dimensional gases of hard elastic spheres *Phys. Rev.* **50** 955

[80] Girardeau M 1960 Relationship between systems of impenetrable bosons and fermions in one dimension *J. Math. Phys.* **1** 516–23

[81] Ouvry S and Polychronakos A P 2009 The Lieb-Liniger model in the infinite coupling constant limit *J. Phys. A: Math. Theor.* **42** 275302

[82] Sklyanin E K 1980 Quantum version of the method of inverse scattering problem *Zap. Nauchn. Semin.* **95** 55–128

[83] Klassen T R and Melzer E 1990 Purely elastic scattering theories and their ultraviolet limits *Nucl. Phys.* B **338** 485–528

[84] Wilson J M, Malvania N, Le Y, Zhang Y, Rigol M and Weiss D S 2020 Observation of dynamical fermionization *Science* **367** 1461–4

[85] Roy A, Schuricht D, Hauschild J, Pollmann F and Saleur H 2021 The quantum sine-Gordon model with quantum circuits *Nucl. Phys.* B **968** 115445

[86] Wybo E, Knap M and Bastianello A 2022 Quantum sine-Gordon dynamics in coupled spin chains *Phys. Rev.* B **106** 075102

[87] Lencsés M, Mussardo G and Takács G 2023 Quantum integrability vs experiments: correlation functions and dynamical structure factors. *arXiv preprint arXiv: 2303.16556*

[88] Maruyoshi K, Okuda T, Pedersen J W, Suzuki R, Yamazaki M and Yoshida Y 2023 Conserved charges in the quantum simulation of integrable spin chains *J. Phys. A: Math. Theor.* **56** 165301

[89] Kulkarni G V 2020 Asymptotic analysis of the form-factors of quantum spin chains *PhD thesis* Bourgogne U.

[90] Coleman S R and Thun H J 1978 On the prosaic origin of the double poles in the sine-Gordon S matrix *Commun. Math. Phys.* **61** 31

[91] Smirnov F A 1990 Reductions of the sine-Gordon model as a perturbation of minimal models of conformal field theory *Nucl. Phys.* B **337** 156–80

[92] Ahn C, Delfino G and Mussardo G 1993 Mapping between the sinh-Gordon and ising models *Phys. Lett.* B **317** 573–80

[93] Chew G F and Frautschi S C 1961 Principle of Equivalence for all strongly interacting particles within the S matrix framework *Phys. Rev. Lett.* **7** 394–7

[94] Karowski M and Thun H J 1977 Complete S matrix of the massive Thirring model *Nucl. Phys.* B **130** 295–308

[95] Zamolodchikov A B 1977 Quantum sine-Gordon model. The total S matrix Report no. ITEP-12-1977 (Microfiche at Fermilab)

[96] Ghoshal S and Zamolodchikov A 1994 Boundary s matrix and boundary state in two-dimensional integrable quantum field theory *Int. J. Mod. Phys.* A **9** 3841–85

[97] Faddeev L D 1996 How algebraic Bethe ansatz works for integrable model *Les Houches School of Physics: Astrophysical Sources of Gravitational Radiation* (Les Houche: School of Physics) 149–219 pp

[98] Kassel C 1995 *Quantum Groups* (New York: Springer Science and Business Media)

[99] Faddeev L D and Tirkkonen O 1995 Connections of the Liouville model and XXZ spin chain *Nucl. Phys.* B **453** 647–69

[100] Destri C and de Vega H J 1987 Light cone lattice approach to fermionic theories in 2D: the massive Thirring model *Nucl. Phys.* B **290** 363–91

[101] Destri C and de Vega H J 1989 Light cone lattices and the exact solution of Chiral Fermion and σ models *J. Phys.* A **22** 1329

[102] Fendley P and Saleur H 1994 Deriving boundary S matrices *Nucl. Phys.* B **428** 681–93

[103] Kirillov A N and Yu Reshetikhin N 1987 Exact solution of the integrable XXZ Heisenberg model with arbitrary spin. I. The ground state and the excitation spectrum *J. Phys.* A **20** 1565–85

[104] Doikou A and Nepomechie R I 1998 Discrete symmetries and S matrix of the XXZ chain *J. Phys.* A **31** L621–8

[105] Doikou A and Nepomechie R I 1999 Direct calculation of breather S matrices *J. Phys.* A **32** 3663–80

[106] T $\overline{\text{T}}$ Y 2023 Ttbar-deformation: a lattice approach *Symmetry* **15** 2212

[107] Korepin V E 1979 Direct calculation of the S-matrix in the massive Thirring model *Theor. Math. Phys.* **41** 953–67

[108] Andrei N and Destri C 1984 Dynamical symmetry breaking and fractionization in a new integrable model *Nucl. Phys.* B **231** 445–80

[109] Miranda E 2003 Introduction to bosonization *Braz. J. Phys.* **33** 3–35

[110] Murugan J and Nastase H 2019 One-dimensional bosonization and the SYK model *JHEP* **08** 117

[111] Isaev A P 2022 *Lectures on Quantum Groups and Yang-Baxter equations* **6**

[112] Crampé N, Gaboriaud J, Vinet L and Zaimi M 2020 Revisiting the Askey–Wilson algebra with the universal r-matrix of *J. Phys. A: Math. Theor.* **53** 05LT01

[113] Delius G W 1995 Exact S matrices with affine quantum group symmetry *Nucl. Phys.* B **451** 445–68

[114] Doikou A, Evangelisti S, Feverati G and Karaiskos N 2010 Introduction to quantum integrability *Int. J. Mod. Phys.* A **25** 3307–51

[115] Reshetikhin N and Smirnov F 1990 Hidden quantum group symmetry and integrable perturbations of conformal field theories *Commun. Math. Phys.* **131** 157–78

[116] LeClair A and Vafa C 1993 Quantum affine symmetry as generalized supersymmetry *Nucl. Phys.* B **401** 413–54

[117] Jackiw R and Rebbi C 1976 Solitons with Fermion number 1/2 *Phys. Rev.* D **13** 3398–409

[118] Khoroshkin S M and Tolstoy V N 1993 Twisting of quantum (super-) algebras *Connection of Drinfeld's and Cartan-Weyl Realizations for Quantum Affine Algebras, Proc. Int. Symp. on Mathematical Physics* (Clausthal: Arnold Sommerfeld Institute) 42–54 pp

[119] Ding J, Pakuliak S and Khoroshkin S 2000 Factorization of the universal r-matrix for *Theor. Math. Phys.* **124** 1007–37

[120] Zhang Y-Z and Gould M D 1994 Quantum affine algebra and universal R matrix with spectral parameter *Lett. Math. Phys.* **31** 101–10

IOP Publishing

Integrability using the Sine-Gordon and Thirring Duality
An introductory course
Alessandro Torrielli

Chapter 4

The Thirring model

4.1 Fermions in the game

We are now ready to introduce the second main character of this book, the Thirring model. The Lagrangian of the massive Thirring model is given by

$$L_T = \bar{\psi}_T(i\gamma^\mu \partial_\mu - m_T)\psi_T - \frac{g}{2}\bar{\psi}_T\gamma^\mu\psi_T\,\bar{\psi}_T\gamma_\mu\psi_T, \tag{4.1.1}$$

in terms of a Dirac fermion field ψ_T in $1 + 1$ dimensions, with the associated Clifford-algebra gamma matrices [1]:

$$\gamma^0 = \sigma_1, \qquad \gamma^1 = i\sigma_2 \quad \longrightarrow \quad \gamma^3 = \gamma^0\gamma^1 = -\sigma_3 \tag{4.1.2}$$

(in terms of the canonical Pauli matrices). We will often simply call this the 'Thirring model', but we should more carefully always add the attribute of 'massive' because by 'Thirring model' some authors mean to denote the massless version.

The fermion field has engineering mass-dimension $\frac{1}{2}$, so the four-fermion term is (superficially) marginal in $1 + 1$ dimensions, with g a-dimensional. The coupling g is not renormalised (with the cautionary discussion reported for instance under formula (1.5) of [2]), but the mass is. The model is known to be quantum-mechanically well defined for $g > -\pi$ [2]. The limit $g \to -\pi^+$ is the limit of *strong repulsion* [3]: the name will make a lot of sense from the point of view of the duality with Sine-Gordon—see equation (5.1.1), which implies $\beta^2 \to \infty$ in this limit. For $g > -\pi$ the theory is renormalisable—see for instance [4, 5].

The massive theory is integrable [6, 7] and $\mathfrak{u}(1)$-invariant under the global phase shift

$$\psi_T \to e^{i\delta}\psi_T, \qquad \delta \in \mathbb{R}. \tag{4.1.3}$$

The associated conserved charge will be dual to the topological charge of the Sine-Gordon solitons/antisolitons. In fact, under the duality conjectured by Coleman the elementary Thirring fermion should behave like the soliton in Sine-Gordon (see the

doi:10.1088/978-0-7503-5899-6ch4

next section). Quantum integrability follows partly from the duality with Sine-Gordon, but more intrinsically from the relationship with the XXZ spin-chain (see one of the exercises in the next section). As in the case of the Sine-Gordon model, the XXZ chain can be thought of as a lattice regularisation of the Thirring model [4].

Setting $m_T = 0$ gives the massless Thirring model, which is also invariant under chiral transformations—rotating independently the chiral spinor components $\frac{1 \pm \gamma_3}{2} \psi_T$. In addition, the massless model is scale invariant. In the context of condensed-matter theory, the massless Thirring model is known as *Tomonaga–Luttinger (model* or *liquid)*—see for instance [8–11], where some of the physical applications are also discussed. It is in this context that one encounters the technique of bosonisation. A detailed study of the massless model with explicit formulas for its correlations functions is reported in [12], quoted by [2] as having been particularly influential.

4.2 A small snapshot of the 1 + 1-dimensional particle world

So far we have described a great deal of the properties of particles and their scattering in integrable $1 + 1$ dimensional S-matrix theory, and marvelled at the possibility of obtaining exact solutions to problems which in four dimensions have always appeared out of reach. Rarely have we stopped to wonder what the particles that we are scattering in this $1 + 1$-dimensional world of ours really are. In this section, as a preamble to the next section where Coleman's duality will be described, we attempt to gather a few more thoughts and perhaps take a closer look at this rather bizarre environment. The interested reader is invited to start a personal exploration, perhaps beginning with [13].

4.2.1 Representations for fields

Let us begin by the notion that the fields of a theory are characterised by having indices transforming according to an irreducible (for our purposes, finite-dimensional and not necessarily unitary) representation of the Lorentz group. If we restrict to the orthochronous proper component, then we can see that in two dimensions such a component is itself an Abelian group, consisting of the transformations (boosts) of the form

$$\Lambda_\lambda = \begin{pmatrix} \cosh \lambda & \sinh \lambda \\ \sinh \lambda & \cosh \lambda \end{pmatrix} \in SO(1, 1), \qquad \lambda \in \mathbb{R}. \tag{4.2.1}$$

It is easy to see that the action of this matrix on a Lorentz vector $\begin{pmatrix} \cosh \theta \\ \sinh \theta \end{pmatrix}$ is to shift $\theta \to \theta + \lambda$. For a small parameter λ, the infinitesimal transformation is generated by the matrix

$$b = \begin{pmatrix} 0 & 1 \\ 1 & 0 \end{pmatrix}. \tag{4.2.2}$$

The Lie algebra is one-dimensional.

According to a known theorem, all irreducible representations of $SO(1, 1)$ are one-dimensional, therefore we should take this into account. This is before specifying any particular theory and only considering fields transforming as irreducible representations of the Lorentz group. The traditional fields that we always tend to write, e.g., the scalar $\phi(x, t)$, the vector $A_\mu(x, t)$, $\mu = 0, 1$ and the Dirac spinor $[\psi_T]_\alpha(x, t)$, $\alpha = 1, 2$, do indeed all reduce to one-dimensional irreducible representations: in a suitable basis, we can write these one-dimensional irreps as

$$\phi, \qquad A_0 + A_1, \qquad A_0 - A_1, \qquad \begin{pmatrix} \chi \\ \chi \end{pmatrix}, \qquad \begin{pmatrix} \chi \\ -\chi \end{pmatrix}. \tag{4.2.3}$$

Let us try to embed this into the representation theory of $SO(1, 1)$ (which is isomorphic to $\mathbb{R}$ when the latter is regarded as a Lie group with the additive operation). The irreducible representations are all one-dimensional and come in a continuous one-parameter family, labelled by a real number s. The vector space carrying the representation is $\mathbb{R}$ in all cases, and to any element (4.2.1) of $SO(1, 1)$ we associate the map

$$\rho_s: SO(1, 1) \rightarrow GL(1), \qquad \rho_s(\Lambda_\lambda)z = e^{s\lambda}z, \qquad z \in \mathbb{R}. \tag{4.2.4}$$

We can equivalently express this by introducing an object x_s whose transformation property under the single generator at our disposal (the boost b) is as follows:

$$x_s \rightarrow e^{s\lambda}x_s \qquad \text{as} \qquad \theta \rightarrow \theta + \lambda, \qquad \text{where} \qquad x_s \equiv e^{s\theta}. \tag{4.2.5}$$

The label s is called the *Lorentz spin*. It is continuous as opposed to the ordinary (sometimes called *intrinsic* [13]) spin in $3 + 1$ dimension—where it is instead related to $SU(2)$ and famously only acquires discrete values. It is easy to now embed the one-dimensional representations (4.2.3) into the classification (4.2.4): respectively, in the order as they appear in (4.2.3), we have Lorentz spin $s = 0, 1, -1, \frac{1}{2}, -\frac{1}{2}$. Continuous values for the Lorentz spin s can be considered. A very early mention of such 'interpolation' is already made, for example, in [12].

Notice that even though all the irreducible representations are one-dimensional, it is still useful (rather essential in fact) to assemble a multiplicity of irreducible representations into one single field, whose indices therefore will necessarily transform under a reducible representation of $SO(1, 1)$. This is done to better display the extended symmetries relating the different irreducible components in the chosen theory, and/or to write the mutual interactions in a compact and geometric way. In the case of the Thirring model, which is part of our primary interest, the Thirring fermion is assembled using $s = \frac{1}{2}$ and $s = -\frac{1}{2}$ irreducible representations by construction. In addition, by construction the Sine-Gordon field φ is built out of the $s = 0$ irreducible representation.

Having assigned the spin to fields, we have a way to assign the statistics to the field operators. As is remarked in the footnote on page 187 of [14], in a local quantum field theory in two dimensions we (still) have that (under specific assumptions) half-integer values of s are associated with anti-commuting fields— i.e., fields displaying a fermionic statistics—and integer values of s with commuting fields—i.e., fields displaying a bosonic statistics. This is part of (the $1 + 1$-dimensional version of) the spin-statistics theorem.

It is important that we say 'part of' the theorem because in $1 + 1$ dimensions the situation is rather more complex than in $3 + 1$ dimensions. This is precisely connected with the considerations that we have made above about the representations of the Lorentz group in $1 + 1$ dimensions and to the peculiar properties of the $1 + 1$-dimensional Minkowski space. The treatment that clarifies the mathematical sense in which this statement has to be understood is found in [15]. We shall try to summarise some points from [15] to the best of our ability; however, we recommend the reader to consult the paper independently for a most beautiful and gratifying reading.

It is shown in [15] that if one constructs a quantum field theory out of a certain number of basic ('primary') local fields Φ_α, and if one encodes the mutual relative statistics of such fields into a *statistics matrix* $R^{\gamma\delta}_{\alpha\beta}(x, y)$ as in

$$\Phi_\alpha(x)\,\Phi_\beta(y) = R^{\gamma\delta}_{\alpha\beta}(x, y)\,\Phi_\gamma(y)\,\Phi_\delta(x) \qquad (4.2.6)$$

(for space-like separated x and y), then there are a number of remarkable facts that occur in $1 + 1$ dimensions and which do not occur in $3 + 1$ dimensions. First of all, the entries of the matrix R are not necessarily always just signs (as it is for bosonic and fermionic fields) but can also be given by more general statistical factors. These 'exotic' statistics are realised in plenty of models and concrete instances. To see precisely how this comes about, a relation on R is derived in [15] based on a series of assumptions about the structure of the theory that can be found explicitly stated in [15] and which we leave to the reader for an interesting individual exploration. The relation derived in this way reads in matrix form

$$R\big(\text{sign}(x_1 - y_1)\big)\,R\big(-\text{sign}(x_1 - y_1)\big) = \mathbb{1}, \qquad (4.2.7)$$

with x_1 and y_1 the spatial components of x and y, respectively. It is in fact argued that R can only at most depend on $\text{sign}(x_1 - y_1)$ in $1 + 1$ dimensions. The analogue of the condition (4.2.7) in $3 + 1$ dimensions reduces drastically. This is because $\text{sign}(x_1 - y_1)$, when x_1 and y_1 belong to space-like separated points, is Lorentz-invariant in $1 + 1$ dimensions.

Exercise [20 minutes' work]: Consider the quantity $\sigma_F \equiv \text{sign}(x_1 - y_1)$ in $1 + 1$ dimensions, for space-like separated points x and y. Feel free to set y to be the origin for simplicity. Clearly σ_F determines in which of the two disconnected wedges of the exterior of the light-cone we are sitting in, and therefore it has to be Lorentz-invariant. Prove that indeed it is by applying a $1 + 1$-dimensional Lorentz transformation and using simple inequalities of elementary functions.

In $3 + 1$ dimensions R cannot depend on the coordinates at all, hence it has to be a constant, and therefore

$$R^2 = \mathbb{1}. \qquad (4.2.8)$$

This restricts the statistics matrix to the point where only bosons and fermions are compatible with (4.2.7) in $3 + 1$ dimensions. On the contrary, a much richer array of possibilities holds in $1 + 1$ dimensions where (4.2.7) allows for a much wider set of solutions (sometimes referred to as *braid-group statistics*).

But there is more. Once again based on a series of assumptions related to the structural features of the theory and to the positivity of the metric in the physical Hilbert space, by studying the transformation properties under Lorentz boosts of the basic field Φ_α and their two-point correlator functions, [15] concludes that the matrix R must be such that

$$\sum_\rho e^{2\pi i s_\alpha}\, \mu_\alpha^\gamma\, \bar{\mu}_\kappa^\rho\, R_{\gamma\alpha}^{\kappa\rho}(\sigma)\, G_\rho\big((0, y_1), (0, x_1)\big) \quad \text{be positive definite,} \qquad (4.2.9)$$

where a bar - denotes complex conjugation. Here

$$\delta_{\alpha\beta}\, G_\alpha(x, y) = \langle \Phi_\alpha(x)|\Phi_\beta(y)\rangle, \qquad (4.2.10)$$

s_α is the spin of the field Φ_α, and the matrix μ is such that

$$\Phi_\alpha^*(x) = \mu_\alpha^\beta\, \Phi_\beta(x), \qquad (4.2.11)$$

* being a certain automorphism of the field algebra. At this point the argument is that in $3 + 1$ dimensions the spin s_α is the intrinsic spin, and therefore only integer or semi-integer. We already know that $R^2 = 1$ and (4.2.9) links the spin and the statistics in the usual way, which is familiar from $3 + 1$-dimensional quantum field theory. On the other hand, in $1 + 1$ dimensions one has a host of possibilities for s_α and for R. There is a link that establishes a connection between the spin and the statistics, which is given by (4.2.9), but one needs to work out case by case what this link is by solving (4.2.9) for the theory at hand.

Nevertheless, it is worthwhile to remark that we can even have simplifications in $1 + 1$ dimensions. If we admit a situation with a single basic field of Lorentz spin s, then the condition (4.2.9) reduces to

$$e^{2\pi i s}\bar{\mu}\mu\, R(\sigma)\, G\big((0, y_1), (0, x_1)\big) \geqslant 0. \qquad (4.2.12)$$

The quantity μ is just a number, in fact now $\bar{\mu}\mu = 1$ [15]. Given that G is positive because of the positivity of the Hilbert-space metric, we conclude that

$$e^{2\pi i s}R(\sigma) \geqslant 0. \qquad (4.2.13)$$

If we further restrict to integer/semi-integer s, and considering what the other relationships listed in [15], which R has to satisfy, reduce to for one single field, we can see that (4.2.13) effectively reduces to the standard commutation/anticommutation relation, respectively.

For a discussion of the implications of this result in the context of conformal field theory, the reader is recommended to consult, for instance, [16].

We conclude this subsection with a very amusing fact. The associativity of the algebra of fields, generated by the basic fields Φ_α, is proven in [15] to be equivalent to the requirement that $R^{\gamma\delta}_{\alpha\beta}(x, y)$ be a solution of the Yang–Baxter equation, with the argument $\text{sign}(x_1 - y_1)$ of R playing the role of $\theta_1 - \theta_2$ in the standard form in which we always write the integrable relativistic S-matrix.

4.2.2 Representations for particles

The space of states (particles) where the fields act as quantum operators carries a *unitary* irreducible representation of the *Poincare' group*. Such a representation is necessarily infinite-dimensional, as can be concretely seen by the fact that it has to accommodate arbitrary values of the momentum $|p_0, p_1\rangle$.

The peculiarity of the irreducible representations in $1 + 1$ dimensions, however, results in the fact that it is possible to have a description of the behaviour of the same particle using fields of different statistics. This is the key to the duality that we are about to describe in the next section, and which we will present in detail. We shall see that there is still a direct connection between the statistics of a field (operator) and of its elementary modes (perturbative excitations). We shall also see that one is able to describe the behaviour of the very same modes as nonelementary (and in fact rather complicated) excitations of a field of a *different* statistics. In particular, the behaviour of the elementary excitations of a fermion field—which are fermionic excitations—can be described in terms of a bosonic field.

This is connected with the intuition that we have already gained about the way in which the interaction can alter, in a nonperturbative way, the behaviour of a particle as regards to the statistics. This is due to the fact that the Bethe wave-function has the exact nonperturbative S-matrix as an ingredient, in addition to the operators creating the particles. The value of $S(0) = -1$ can, for example, enforce an exclusion principle for bosons from the nonperturbative account of the interaction, as we have seen in the case of the Lieb–Liniger model.

References

[1] Tong D *Gauge Theory–Part III Cambridge Lecture Notes.* https://www.damtp.cam.ac.uk/user/tong/gaugetheory.html (accessed March 25 2024)

[2] Coleman S R 1975 The quantum sine-Gordon equation as the massive Thirring model *Phys. Rev.* D **11** 2088

[3] Korepin V E 1980 The mass spectrum and the s matrix of the massive thirring model in the repulsive case *Commun. Math. Phys.* **76** 165–76

[4] Luther A 1976 Eigenvalue spectrum of interacting massive fermions in one-dimension *Phys. Rev.* B **14** 2153–9

[5] Luscher M 1976 Dynamical charges in the quantized, renormalized massive Thirring model *Nucl. Phys.* B **117** 475–92

[6] Mikhailov A V 1976 About integrability of two-dimensional Thirring model *Pisma Zh. Eksp. Teor. Fiz* **23** 356–8

[7] Kuznetsov E A and Mikhailov A V 1977 On complete integrability of two-dimensional classical Thirring model *Teor. Mat. Fiz.* **30** 303–14

[8] Mattis D C and Lieb E H 2004 Exact solution of a many-Fermion system and its associated boson field *Condensed Matter Physics and Exactly Soluble Models: Selecta of Elliott H. Lieb* (Berlin: Springer) 645–53 pp

[9] Haldane F D M 1981 'Luttinger liquid theory'of one-dimensional quantum fluids. i. properties of the Luttinger model and their extension to the general 1d interacting spinless fermi gas *J. Phys. C: Solid State Phys.* **14** 2585

[10] Fradkin E 1991 Field theories of condensed matter (Cambridge: Cambridge Univ. Press)

[11] Degiovanni P, Mélin R and Chaubet Ch 1998 Conformal field theory approach to gapless 1d fermion systems and application to the edge excitations of $\nu = 1/(2p + 1)$ quantum hall sequences *Theor. Math. Phys.* **117** 1113–81

[12] Klaiber B 1968 The Thirring model *Lect. Theor. Phys.* A **10** 141–76

[13] Swieca J A 1977 Solitons and confinement *Fortsch. Phys.* **25** 303–26

[14] Green M B, Schwarz J H and Witten E 1988 *Superstring Theory. Vol 1: Introduction* **7** Cambridge Monographs on Mathematical Physics (Cambridge: Cambridge Univ. Press)

[15] Fröhlich J 1988 Statistics of fields, the Yang–Baxter equation, and the theory of knots and links *Nonperturbative Quantum Field Theory* (Singapore: World Scientific) 71–100 pp

[16] Gaberdiel M R 2000 An introduction to conformal field theory *Rep. Prog. Phys.* **63** 607

IOP Publishing

Integrability using the Sine-Gordon and Thirring Duality
An introductory course
Alessandro Torrielli

Chapter 5

Duality between Sine-Gordon and Thirring

5.1 Coleman's argument

We now come to the core of the duality, of which we have already disseminated a few hints. It was stated most effectively by Coleman [1] and did generate a massive amount of activity, highlighting some of the most striking features of (integrable) two-dimensional quantum field theories.

The statement that Coleman argued was as follows: there is a relationship between the two theories (later we will see that this involves specific sectors thereof), established by

1. A relation between the parameters:

$$\frac{4\pi}{\beta^2} = 1 + \frac{g}{\pi}, \qquad \beta^2 < 8\pi, \tag{5.1.1}$$

$\beta^2 > 0$ being consistent with $g > -\pi$. This is in line with the non-renormalisation of either couplings appearing in (5.1.1). In particular, $\beta^2 = 4\pi$ corresponds to free fermions $g = 0$ (see our later discussion of [2]). It is noteworthy that small β corresponds to g going to infinity (an instance of weak-strong duality, akin to S-duality). The relation (5.1.1) is also consistent with the fact that $\beta^2 = 4\pi$ signals the passage between attractive and repulsive regimes for the soliton (resp. fermion) interaction.

2. A relation between specific (combinations of) fields:

$$-\frac{\beta}{2\pi}\epsilon^{\mu\nu}\partial_\nu\phi = \bar{\psi}_T\gamma^\mu\psi_T, \qquad \frac{m^2}{\beta^2}\cos\beta\phi = -Zm_T\bar{\psi}_T\psi_T, \tag{5.1.2}$$

to be understood in the sense of perturbative calculations of correlation functions, with Z incorporating the regularisation. The second formula in (5.1.2) is then refined to [1, 3–5]

$$-Zm_T\bar{\psi}_T\left(\frac{1 \mp \gamma_3}{2}\right)\psi_T = \frac{m^2}{\beta^2}e^{\pm i\beta\phi}. \qquad (5.1.3)$$

Note that Coleman used a very specific perturbation theory[1], performed around vanishing mass, with a delicate treatment of the severe infrared issues presented by a massless scalar field in two dimensions (see for example [7]). The duality eventually transcends the particular perturbative analysis and regularisation scheme used, and is in fact directly displayed using the respective exact expressions for the S-matrices.

It is important to note (see also the Conclusions) that the 'dictionary', as originally stated by Coleman, involves only relations between objects with zero total topological/u(1) charge.

Perturbative checks at the level of the S-matrix are performed for instance in [8]— in the case of Thirring one uses standard (small g) perturbation theory (more on this in the next section), while in the case of the solitons one can resort to semiclassics [9]. These perturbative-type checks of the exact formula are in rather distant regions of validity. In fact, the semiclassical regime corresponds to small ξ because by looking at the $\frac{1}{\hbar}\int d^2x\frac{m^2}{\beta^2}\cos\beta\phi$ term we see that the dimensionless parameter in Sine-Gordon is $\hbar\beta^2$. Therefore, the expansion in β^2 is also an expansion in $\hbar$.

It is worth noticing that the Sine-Gordon singularity highlighted by Coleman, occurring as β^2 reaches 8π from below, has a correspondence in Thirring where g reaches $-\frac{\pi}{2}$ from above. When regarded as a perturbation of the massless model, the Thirring mass-term perturbation is relevant for $g > -\frac{\pi}{2}$ [2]. Coleman [1] provides an argument based on the energy density to demonstrate that it becomes unbounded from below under this critical value. The singularity was later argued to only be apparent [3–5].

Although these days, after the advent of AdS/CFT, we are in some sense used to the most dazzling dualities between theories who do not even remotely resemble one another, at the time when Coleman proposed this particular duality he was in fact pointing the finger on a remarkable property of $1 + 1$-dimensional theories. In particular, the rearrangement of some of the degrees of freedom of one theory into the semblance of those of the other is a highly nontrivial and nonperturbative phenomenon.

Mandelstam [10] (see also [11]) has constructed the quantum soliton creation and annihilation operators directly within Sine-Gordon, in terms of normal-ordered exponentials of the (integral of the) bosonic field and its derivatives. These operators have spin 0 and are bosonic (commuting). They are characterised by an S-matrix satisfying $S(0) = -1$, whence the fermionic behaviour of the Bethe states.

---

[1] Effectively employing what is known as conformal perturbation theory, in which the emphasis is drawn to correlation functions of objects with a well-defined scaling dimension—see also [6]. For the conventions on the mass regulator employed in the Gaussian conformal field theory (CFT), we refer for example to appendix A in [2].

As we will shortly review, Mandelstam has also shown how to construct operators satisfying anticommutation relations using the Sine-Gordon bosonic field. These naturally behave as the Thirring fermions. We refer to [10, 11] for a discussion of the subtleties associated with this construction in the context of the axiomatic approach to quantum field theory—see also [12]. We will discuss some more on this point in the next section.

The mass relationship depends on the renormalisation scheme used in the two theories, but it is natural to adjust the two schemes such that the finite parts satisfy [13]

$$m_{T,r} = m_{sol}, \qquad (5.1.4)$$

where $m_{T,r}$ is the renormalised fermion mass.

The duality finds a deep realisation within the common integrable structure, ultimately controlled by the underlying quantum group. This is manifested in the Bethe ansatz formulation of the spectral problem, which was studied for instance in [14, 15]. For a review of the method of the Bethe ansatz to solve integrable systems, we refer to [16–19]—see also the next section.

5.2 Project

Project [8 weeks' work]: Study the papers [14, 15]. This will show how integrability allows us to construct an exact analogue of the procedure of filling the Dirac sea for fermionic quantum field theories in a fully interacting theory as opposed to free fields (perturbatively). The conservation of the particle number and the use of the Bethe wave-function is crucial to that purpose. An interesting addendum to the project could be to reproduce the calculations of [14, 15]: in these references interesting considerations are made concerning renormalisation from the perspective of the Bethe ansatz.

A fact that we have already remarked upon in the context of Coleman's duality is that the 'free-fermion' point of the Thirring model is dual to Sine-Gordon at the special value $\beta = \sqrt{4\pi}$, although more insight will be given in the following when we discuss the analysis by Klassen and Melzer. If we restrict ourselves to the massless case, then we can make use, even for values of β other than $\sqrt{4\pi}$, of the technique of *bosonisation*. We recommend the excellent review [20] for a pedagogical treatment of bosonisation—see also [6, 21, 22] for a discussion of some of the subtleties related to this point.

Literature search [unspecified amount of time]: Explore the literature on bosonisation [22] as a way of relating a free boson to an interacting massless fermion. Study the sections of [22] describing chiral bosons and the bosonisation dictionary. As explained on page 148 of [23], considering 1 + 1 dimensional systems introduces a number of special features that have an impact on the formulation and the implementation of the bosonisation procedure. Higher-dimensional versions of bosonisation have been studied—see for instance [24–26]. Explore the literature of Jordan–Wigner transformations—starting for instance from [27, 28].

Literature search [unspecified amount of time]: Explore the literature of massless integrable scattering, taking the moves from [29], see also [30]. You will discover the issue of left and right movers, and the theory of massless flows. See in particular [31, 32]. The literature on massless integrable scattering is enormous and reaches all the way to the most recent applications in holography. A start in this direction can be found in the supplement.

5.3 Mandelstam's construction

Mandelstam [10] has shown how to construct soliton creation and annihilation operators, using the bosonic sine-Gordon field as a dynamical variable. He has also shown how to construct operators in such a way that the resulting expressions satisfy the anticommutation relations of the massive Thirring field, which is a fermion. Along the way, some of the relations that we have described in the previous section as being part of the 'dictionary' of the duality have then been verified explicitly by Mandelstam using these expressions.

Let us focus on the operators that are associated with the Thirring fermion and begin by writing their expressions, verifying one of the relations as a simple example. Mandelstam defines two operators, which we have here renamed as $\psi_\pm$ (they are called ψ_2 and ψ_1, respectively, in [10]):

$$\psi_\pm(x,\,t) \propto \exp\left[-\frac{2\pi i}{\beta}\int_{-\infty}^{x} d\xi\,\partial_t\phi(\xi,\,t) \pm \frac{ik\beta}{2}\phi(x,\,t)\right], \qquad (5.3.1)$$

with k a real number, which we introduce and which we keep for the moment for our convenience. All the expressions that we write here—in (5.3.1) and also in the remainder of this section—are understood as being properly normal-ordered, with the normal ordering being denoted wherever it is done explicitly by the traditional symbol $::$, and regulated wherever necessary. There is also a proportionality constant in (5.3.1), which we, however, omit as not directly essential to our exercise (the precise value of the proportionality constant can be found in [10]). Ignoring these subtleties means that we shall proceed rather formally and take some shortcuts— referring to the original [10] for a more rigorous treatment.

If we suppress the time dependence for the remainder of the section and denote the exponent of $\psi_\pm$ in (5.3.1) by

$$-\frac{2\pi i}{\beta}\int_{-\infty}^{x} d\xi\,\dot\phi(\xi) \pm \frac{ik\beta}{2}\phi(x) \equiv A_\pm, \qquad (5.3.2)$$

then we can consider as an example

$$\psi_+(x)\psi_+(y) =: e^{A_+(x)} :: e^{A_+(y)}: \qquad x \neq y. \qquad (5.3.3)$$

Let us begin with setting $k = 1$. Let us ignore for the moment another issue, which might concern the reader who is wondering about the well-definiteness of the exponential operator. Setting $k = 1$ appears to naively contrast with the periodicity of the Sine-Gordon field $\phi \sim \phi + \frac{2\pi}{\beta}$. In a later chapter, when discussing the paper by Klassen and Melzer, we shall return to this point. For now let us simply proceed formally, in anticipation of a later explanation.

Let us consider that the normal ordering is implemented by accumulating all the creation operators to the left of all the annihilation operators. The field, its derivatives and all the linear functionals $F(\phi, \partial_\mu\phi)$ thereof will schematically split as

$$F = F^{(+)} + F^{(-)}, \tag{5.3.4}$$

if we denote by $F^{(\pm)}$ the creation part, resp. the annihilation part. We then use a formula from [10], which says that if $[F^{(-)}, G^{(+)}]$ is a 'constant' (namely, not an operator but possibly a function), then

$$:e^F :: e^G := e^{[F^{(-)}, G^{(+)}]}: e^{F+G}: . \tag{5.3.5}$$

This follows from Campbell–Baker–Hausdorff's formula: $:e^F :: e^G := e^{F^{(+)}}e^{F^{(-)}}e^{G^{(+)}}e^{G^{(-)}}$ $= e^{[F^{(-)}, G^{(+)}]}e^{F^{(+)}}e^{G^{(+)}}e^{F^{(-)}}e^{G^{(-)}} = e^{[F^{(-)}, G^{(+)}]}e^{(F+G)^{(+)}}e^{(F+G)^{(-)}}$, if $[F^{(-)}, G^{(+)}]$ is a constant (and given that parts labeled by plus commute among themselves, and parts labeled by minus commute amongst themselves). This proves (5.3.5). This means that if $[G^{(-)}, F^{(+)}]$ is a constant,

$$:e^G :: e^F := e^{[G^{(-)}, F^{(+)}]}: e^{F+G}: . \tag{5.3.6}$$

and therefore

$$:e^G :: e^F := e^{[G^{(-)}, F^{(+)}]} e^{-[F^{(-)}, G^{(+)}]}: e^F :: e^G := e^{[G, F]}: e^F :: e^G: , \tag{5.3.7}$$

where we have assumed that creation commutes with creation and annihilation commutes with annihilation because our fields will be canonical. We wish to apply this to our situation, where indeed the relevant commutators of the $(+)$ and $(-)$ parts will always be constant. Let us focus on the commutator part: by plugging in the explicit expressions, and using the fact that the field φ commute with itself for space-like separations, we can be led to

$$[A_+(x), A_+(y)] = -\pi \int_{-\infty}^x d\xi \, [\dot\phi(\xi), \phi(y)] - \pi \int_{-\infty}^y d\xi \, [\phi(x), \dot\phi(\xi)]. \tag{5.3.8}$$

Now we use the equal-time commutator $[\phi(x), \dot\phi(y)] = i\delta(x - y)$ and realise that:
1. If $x > y$, then the first integral contributes $i\pi$ and the second integral is 0;
2. If $x < y$, then the second integral contributes $-i\pi$ and the first integral is 0;

either way we get $e^{[A_+(x), A_-(y)]} = -1$, and hence

$$\psi_+(x)\psi_+(y) = e^{A_+(x)}e^{A_+(y)} = -\psi_+(y)\psi_+(x), \qquad x \neq y. \tag{5.3.9}$$

We remind that x and y are the spatial locations (because time is suppressed). The operators are therefore shown to be fermionic, i.e., they *anticommute*.

It is clear from our presentation that the limit $x \to y$ in this type of reasoning is particularly delicate and would require a much more careful treatment. We refer to [10] for further analysis.

Let us now look at Mandelstam's operators with the assignment $k = 0$. Such operators commute instead of anti-commuting. We can understand the fact that the operators $\psi_\pm$ carry a topological charge. Mandelstam notices that these operators satisfy the following commutation relations with the bosonic field φ:

$$[\phi(y), \psi_\pm^\dagger(x)] = -\frac{2\pi}{\beta}\psi_\pm^\dagger(x) \qquad y < x, \tag{5.3.10}$$

$$[\phi(y), \psi_\pm^\dagger(x)] = 0 \qquad y > x. \tag{5.3.11}$$

The realisation of this fact is not too difficult if we rely again upon the fact that $\psi_\pm \propto e^{A_\pm}$, with $A_\pm$ defined in (5.3.2). Let us first notice that

$$[:e^{A_\pm(x)}:]^\dagger \phi(y) = e^{(A_\pm^\dagger)^{(-)}(x)} e^{(A_\pm^\dagger)^{(+)}(x)} \phi(y). \tag{5.3.12}$$

Then if we assume that

$$[(A_\pm^\dagger)^{(+)}(x), \phi(y)] = c, \tag{5.3.13}$$

with c a constant (not an operator), then we have

$$
e^{(A_\pm^\dagger)^{(+)}(x)}\phi(y) = \sum_{n=0}^\infty \frac{((A_\pm^\dagger)^{(+)}(x))^n}{n!}\phi(y) = \\
\sum_{n=0}^\infty \frac{1}{n!}(nc((A_\pm^\dagger)^{(+)}(x))^{n-1} + \phi(y)((A_\pm^\dagger)^{(+)}(x))^n),
\tag{5.3.14}
$$

where the last step can be proven by induction. Therefore,

$$e^{(A_\pm^\dagger)^{(+)}(x)}\phi(y) = (c + \phi(y))e^{(A_\pm^\dagger)^{(+)}(x)}. \tag{5.3.15}$$

By the same procedure, if

$$[(A_\pm^\dagger)^{(-)}(x), \phi(y)] = \tilde{c}, \tag{5.3.16}$$

then

$$e^{(A_\pm^\dagger)^{(-)}(x)}e^{(A_\pm^\dagger)^{(+)}(x)}\phi(y) = (c + \tilde{c} + \phi(y))e^{(A_\pm^\dagger)^{(-)}(x)}e^{(A_\pm^\dagger)^{(+)}(x)}. \tag{5.3.17}$$

In our specific case c and $\tilde{c}$ are constant, moreover $c + \tilde{c} = [A_\pm^\dagger(x), \phi(y)]$, which we can easily compute: if $x > y$,

$$[A_\pm^\dagger(x), \phi(y)] = \frac{2\pi i}{\beta}\int_{-\infty}^x d\xi\, [\dot\phi(\xi), \phi(y)] = \frac{2\pi}{\beta}, \tag{5.3.18}$$

while if $y > x$, then clearly $[A_\pm^\dagger(x), \phi(y)] = 0$. If we then focus on $y < x$, we then have

$$[:e^{A_\pm(x)}:]^\dagger \phi(y) = \left(\frac{2\pi}{\beta} + \phi(y)\right)[:e^{A_\pm(x)}:]^\dagger, \tag{5.3.19}$$

which provides a way to arrive at (5.3.10). On the other hand, for $y > x$ we can clearly see that $[:e^{A_\pm(x)}:]^\dagger \phi(y) = \phi(y)[:e^{A_\pm(x)}:]^\dagger$ because, as we have mentioned, in that region $[A_\pm^\dagger(x), \phi(y)] = \frac{2\pi i}{\beta} \int_{-\infty}^{x} d\xi \, [\dot{\phi}(\xi), \phi(y)] = 0$ (the domain does not contain the support of the delta function). We have therefore quickly arrived at (5.3.11).

This means that in the region $y < x$ we have

$$\phi(y)\psi_\pm^\dagger(x)|0\rangle = \psi_\pm^\dagger(x)\left(-\frac{2\pi}{\beta} + \phi(y)\right)|0\rangle, \qquad (5.3.20)$$

while in the region $y > x$ we have

$$\phi(y)\psi_\pm^\dagger(x)|0\rangle = \psi_\pm^\dagger(x)\phi(y)|0\rangle. \qquad (5.3.21)$$

This can be interpreted as the operator $\psi_\pm^\dagger(x)$ 'adding' an amount $-\frac{2\pi}{\beta}$ to the vacuum 'profile' in the region to the left of x, and nothing in the region to the right. Effectively, the action of $\psi_\pm^\dagger(x)$ on the vacuum has created a change in topological charge—if we suitably shift by a constant, then we can use this to 'simulate' the same type of a 'jump' in the topological charge as one observes in the first profile in figure 3.2 as far as the asymptotic values are concerned. We can further interpret this as either creating a Sine-Gordon soliton or annihilating a Sine-Gordon antisoliton—see also a later discussion in the next chapter, at the point where we discuss the paper by Klassen and Melzer.

Notice that the same argument applied to the $k = 1$ case shows that Mandelstam's operators in that case carry a u(1) charge that behaves as the Thirring fermion.

We wish to end this part by remarking that the language of Mandelstam's construction is extremely close to the notion of vertex operators and bosonisation in conformal field theory, which we will see further on when discussing the paper of Klassen and Melzer. The crucial observation comes in how the short-distance singularities are regularised and treated in Mandelstam's analysis: there enters the fact that he is working directly in the massive theory and he is not confining himself to the CFT. The conceptual distinction is further remarked upon, for example, by Klassen and Melzer—please see the discussion above their formula (3.5) in [2].

5.4 Bethe ansatz

One of the alternative ways of testing these ideas is to regard the Sine-Gordon model and the Thirring model through the lens of the Bethe ansatz [18], which is controlled by the same quantum group structure and ends up relying on the same representation for both theories. It is not surprising therefore that the Bethe equations one finds when describing the finite-volume spectrum do correlate (and are both in turn connected with the XXZ spin-chain Bethe ansatz). This gives us an opportunity to explain what the RTT relations (which take their name from the three mathematically symbols, one R-matrix and two monodromy matrices, which appear multiplied in them) are, and to summarise the method of the algebraic Bethe ansatz, as a unified method to treat a variety of different quantum integrable systems. We shall also see how the complication of the Sine-Gordon spectrum (when contrasted, for instance, with the Lieb–Liniger model that we have encountered earlier) forces us towards a nested structure of Bethe states. The primary reason is that, as we have seen, the Sine-Gordon (anti)soliton scattering is characterised by a non-diagonal S-matrix (in the internal space of the topological charge).

The Bethe equations can be constructed by employing the tool of the transfer matrix, which is built as the trace of a string of S-matrices for an ordered sequence of interacting particles. Let us briefly outline the calculation. Consider N relativistic particles on a circle of length L_0. The natural quantisation condition for the momenta can be taken to be (see for example section 3 in [31]):

$$e^{-ip_k L_0} \, \mathrm{tr}_0 T(p_k | p_1, \ldots, p_N) | \psi \rangle \sim | \psi \rangle, \qquad k = 1, \ldots, N, \qquad (5.4.1)$$

where p_i are the momenta of the interacting particles, and the *quantum transfer matrix*, namely the trace over the 0th space of the *quantum monodromy matrix*

$$[T_a^b(p_0 | p_1, \ldots, p_N)]_{c_1 \ldots c_N}^{d_1 \ldots d_N} = \sum_{\{k\}} S_{ac_1}^{d_1 k_1}(\theta_0 - \theta_1) \, S_{k_1 c_2}^{d_2 k_2}(\theta_0 - \theta_2) \ldots S_{k_{N-1} c_N}^{d_N b}(\theta_0 - \theta_N), \quad (5.4.2)$$

appears, S being the S-matrix with the indices associated with the internal degrees of freedom completely spelt out (we will often suppress them hereafter). The trace, as we have just said, is over the 0th space, called *auxiliary*, all the other spaces being called *physical* or *quantum*. The objects appearing here are now the natural quantum versions of the classical monodromy and transfer matrices, where the path-ordered exponential is replaced by an ordered product. The S-matrix itself (or rather its associated R-matrix) is playing the role of a quantum version of the Lax matrix L. In the Hopf-algebra language, where we have seen that it is customary for the indices to denote different spaces in the tensor product, we would write

$$T = \prod_{i=1}^{N} R_{0i}. \qquad (5.4.3)$$

The intuition behind the quantisation condition for the momenta is as follows. When the particle labelled by k goes around the circle of length L_0, it scatters sequentially off all the remaining ones (thanks to the property of factorised scattering). Since the particle has eventually come back to its original position, this must amount to a simple phase shift. We can formalise this by saying that acting on an eigenstate $|\psi\rangle$ of the transfer matrix, and compensating for the global phase shift, should result in an identical wave-function. Notice that we should have excluded the very particle k in the sequence of scatterings. We can, however, formally include it using the fact that $S_{ac}^{db}(0) = \pm \delta_a^d \delta_c^b$ for the particles that we are interested in. A figure which one often finds being drawn in this context is figure 5.1.

The task is then to construct the eigenstates $|\psi\rangle$ of the quantum transfer matrix, thereby realising the diagonalising condition (5.4.1). The algebraic Bethe ansatz precisely achieves this result. The crucial insight is that the monodromy matrix satisfies, by sole virtue of the fact that the S-matrix satisfies the Yang–Baxter equation, a set of relations called RTT, named after their appearance:

$$R_{00'} \, (T_0 \otimes \mathbb{1}_{0'}) \, (\mathbb{1}_0 \otimes T_{0'}) = (\mathbb{1}_0 \otimes T_{0'}) \, (T_0 \otimes \mathbb{1}_{0'}) \, R_{00'}, \qquad (5.4.4)$$

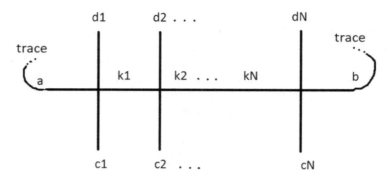

Figure 5.1. A pictorial representation of the monodromy matrix—the auxiliary space being identified as the horizontal line, and being traced over whenever one wishes to obtain the transfer matrix (identifying a with b and summing over).

where a second auxiliary space $0'$ is introduced. The quantum spaces are not explicitly indicated but are common to both T_0 and $T_{0'}$. The auxiliary variables are momenta p_0 and $p_{0'}$, respectively (or, in terms of the rapidities, θ_0 and θ_0').

Exercise [45 minutes' work]: Prove that the Yang–Baxter equation for R combined with (5.4.3) implies the RTT relations. See also the supplement.

Taking the trace $\mathrm{tr}_0 \otimes \mathrm{tr}_{0'}$ on both sides of the RTT relations shows that the trace $\mathrm{tr}_0 T$ commutes with itself at different values of the auxiliary argument. As in the classical case, by expanding in the auxiliary variable (which here plays the role of a spectral parameter for the quantum problem) one generates an infinite set of charges in involution. If we can prove that one of them is the Hamiltonian[2], then these charges are all conserved. To demonstrate how one constructs the eigenvectors, we proceed as follows.

We decompose the monodromy matrix as

$$T_{(\theta_0|\vec{\theta})} = E_{11} \otimes A_{(\theta_0|\vec{\theta})} + E_{12} \otimes B_{(\theta_0|\vec{\theta})} + E_{21} \otimes C_{(\theta_0|\vec{\theta})} + E_{22} \otimes D_{(\theta_0|\vec{\theta})} = \begin{pmatrix} A & B \\ C & D \end{pmatrix}, \quad (5.4.5)$$

having singled out the auxiliary space and having combined into a vector $\vec{\theta}$ all the quantum variables. The trace corresponds to the operator $A + D$. We have used the matrix unities E_{ij}, with all zeroes but 1 in row i and column j, in the auxiliary space. The quantities A, B, C and D are themselves operators that act on the quantum spaces $1, \ldots , N$. The generic eigenvector is built out of *excitations* over a *pseudovacuum* state $|0\rangle$, and is given by the following expression (for M excitations):

---

[2] In ordinary spin-chains the (local) Hamiltonian can be extracted by taking the logarithmic derivative of the transfer matrix evaluated at a special value of the spectral parameter—this is often when all the auxiliary and quantum rapidities (their non-relativistic analogue rather) are set equal [33]. At this point, all the R-matrices, as we have seen before, tend to degenerate to permutation operators by virtue of the braiding unitarity argument. We thank Julius Julius for our conversation on this point.

$$|\beta_1, \ldots, \beta_M\rangle = \prod_{n=1}^{M} B(\beta_n|\vec{\theta})|0\rangle. \tag{5.4.6}$$

The pseudovacuum is dubbed as such because it is not necessarily the ground state of the theory—in fact, in the case of Sine-Gordon it will be the state with N solitons with rapidities $\theta_1, \ldots, \theta_N$. The pseudovacuum has to be a lowest-weight state of the algebra generated by the B and C operators via the RTT relations. Specifically, $|0\rangle$ has to be annihilated by C for all the value of the parameters.

By using the RTT relations, one can prove that (5.4.6) is an eigenvector of the transfer matrix for arbitrary M, thereby diagonalising simultaneously all the commuting charges. By breaking down the RTT relations, one obtains formulas that *morally* read as $[A + D, B] \propto B$: this is in all essence the reason why B qualifies as a creation operator.

Exercise [5 hours' work]: Take the R-matrix associated with the Sine-Gordon model at $\beta^2 = \frac{16\pi}{3}$, and write every entry explicitly using the matrix unities E_{ij}. Derive all the relations involving the operators A, B, C, D which descend from writing the RTT relations in components. You will not find ordinary commutation relations, rather what are known as 'exchange relations', defining an algebra and not a Lie algebra. Such an 'exchange algebra' is actually another way of writing the quantum group controlling the symmetry of the system (the so-called 'RTT presentation' of the quantum group). You can use [16] as a guide—see for instance the formulas (66)–(68) in [16], which are written for the R-matrix of the Heisenberg spin chain. After you have done the first couple of components, you are welcome to write a symbolic computer code that generates them all. A similar exercise is contained in the supplement.

The nesting for Sine-Gordon comes at this stage. It is actually not enough to act with the B operators on the pseudovacuum to create an eigenstate because there is an extra (*unwanted*) term that one needs to cancel. This can be achieved by imposing a first set of Bethe equations, algebraic conditions linking the excitation rapidities β_m to the rapidities θ_i of the initial (sometimes called *frame*) particles. Such equations may be referred to as *level-one* or *auxiliary* Bethe equations. Only after imposing these auxiliary conditions can another set of *momentum-carrying* Bethe equations finally follow from (5.4.1), linking all the rapidities (frame and excitations) to the circle-length L_0 via the transfer matrix eigenvalues. The notion is that the momentum is assigned to each particle, but one can still excite the internal degrees of freedom—by acting with B, one does this very much in the spirit of using the root generators of a Lie algebra to 'create' the various states in a symmetry multiplet.

The energy is one coefficient in the expansion of the transfer matrix eigenvalue, and for Sine-Gordon it will read

$$E_{tot} = \sum_{i=1}^{N} m_{sol} \cosh \tilde{\theta}_i, \tag{5.4.7}$$

with $\{\tilde{\theta}_i\}_{i=1,\ldots,N}$ being each of the possible sets of simultaneous roots of the Bethe equations (quantum configurations). Notice that finding the possible sets of simultaneous Bethe roots $\{\tilde{\theta}_i\}_{i=1,\dots,N}$ involves finding at the same time the corresponding sets of auxiliary roots $\{\tilde{\beta}_j\}_{j=1,\dots,M}$, solutions to the auxiliary Bethe equations. Integrability is again at work, allowing the Bethe ansatz to produce an additive free-particle—type of formula (5.4.7) for the energy (the dynamics being stored in the $\tilde{\theta}_i$ and in m_{sol}). Finding the (auxiliary and momentum-carrying) Bethe roots for arbitrary N, M, L_0 is in general a formidable problem[3].

The explicit set of Bethe equations for Sine-Gordon (including the breathers, which we have ignored so far) can be found for instance in [36]. As we have mentioned earlier, the pseudovacuum is typically made of N solitons, and the excitations correspond to 'flipping' M of them into antisolitons. In the domain of spin-chains these kinds of excitations, obtained by flipping internal degrees of freedom (such as the spins at the various sites of an XXZ spin chain), are often called *magnons*. In Sine-Gordon each B operator creates an antisoliton from a soliton. One needs a different set of operators to create the solitons from the true vacuum in the first place—this extra 'nested' step is missing in the Lieb–Liniger model for example, where one can immediately construct B operators that create the only excitations in the game out of the perturbative vacuum.

The Thirring model Bethe ansatz [14, 15], which is the object of the project that we have outlined in 5.2 and eventually correlates with the Sine-Gordon Bethe ansatz [37], can be seen to be performed using a variant of the algebraic Bethe ansatz, called the *coordinate* Bethe ansatz (which historically predates the former). As we have had the opportunity of highlighting in 5.2, the full detail of the Bethe ansatz construction suited for the duality is highly nontrivial. We again refer to [14, 15, 37] for a complete description.

5.5 Form factors

Another way of probing a theory is by computing form factors. We can do this by using the exact S-matrix for a selection of operators that are easily identifiable and comparing with explicit perturbative computations whenever available.

The n-particle form factor associated with a local operator $\mathcal{O}(x, t)$ is defined as

$$F_{\alpha_1\dots\alpha_n}^{\mathcal{O}}(\theta_1, \dots, \theta_n) = \langle 0|\mathcal{O}(x = 0, t)|\theta_1, \dots, \theta_n\rangle_{\alpha_1\dots\alpha_n}, \qquad (5.5.1)$$

where θ_i is the rapidity of the ith particle *in* state and α_i collectively represents any of the polarisations that the particle has—effectively, the Cartan-subalgebra eigenvalues of a suitable (super)algebra representation.

The axiomatic approach to the study of form factors, due in large measure to Smirnov [38], consists of solving a series of rather general conditions, very much like the ones that determine the exact S-matrix—as we have described them at the beginning of the book. These 'axioms' are also the historical product of abstracting and axiomatising a series of properties that descend from Feynman diagrams.

---

[3] We thank Bogdan Stefański and Yang-Hui He for a discussion on this point. There is an enormous literature on the solutions to the Bethe equations, in fact one might say that this is what integrability ultimately and practically amounts to. For the most recent work that we can think of please see [34, 35].

We present here a summary of these conditions and refer the reader to [8, 39–45] for a more complete treatment.

- *Permutation axiom (or Watson's equation)*

$$F^{\mathcal{O}}_{\alpha_1...\alpha_{j-1}\,\beta_j\,\beta_{j+1}\,\alpha_{j+2}...\alpha_n}(\theta_1, \ldots, \theta_{j-1}, \theta_j, \theta_{j+1}, \theta_{j+2}, \ldots \theta_n)=$$
$$F^{\mathcal{O}}_{\alpha_1...\alpha_{j-1}\,\alpha_j\,\alpha_{j+1}\,\alpha_{j+2}...\alpha_n}(\theta_1, \ldots, \theta_{j-1}, \theta_{j+1}, \theta_j, \theta_{j+2}, \ldots \theta_n)\, S^{\alpha_{j+1}\alpha_j}_{\beta_j\beta_{j+1}}(\theta_j - \theta_{j+1}),$$

(5.5.2)

where the entries of the S-matrix are extracted as we have seen in one of the previous chapters:

$$S^{\alpha\beta}_{\gamma\delta}(\theta_1 - \theta_2) = {}^{out}\langle v_\beta(\theta_2) \otimes v_\alpha(\theta_1)|v_\gamma(\theta_1) \otimes v_\delta(\theta_2)\rangle^{in},$$

(5.5.3)

for a one-particle basis of vectors $\{v_\alpha\}$.

- *Periodicity*

$$F^{\mathcal{O}}_{\alpha_1\alpha_2...\alpha_{n-1}\alpha_n}(\theta_1 + 2i\pi, \theta_2, \ldots \theta_{n-1}, \theta_n) = (-)^\sigma \, F^{\mathcal{O}}_{\alpha_2\alpha_3...\alpha_n\alpha_1}(\theta_2, \theta_3, \ldots \theta_n, \theta_1),$$

where σ is a statistical-type factor, which is graphically often associated with permuting the particle 1 through the operator (see the pictorial representations in section 7 of [45]). This is sometimes phrased in terms of a *semi-locality* index. In certain references, both the permutation and the periodicity axioms are referred to as *Watson's equations*. We refer to section 5.9.1 of [46] for an in-depth explanation of the semi-locality index with physical considerations.

- *Lorentz boost*

$$F^{\mathcal{O}}_{\alpha_1\alpha_2...\alpha_{n-1}\alpha_n}(\theta_1 + \Lambda, \theta_2 + \Lambda, \ldots \theta_{n-1} + \Lambda, \theta_n + \Lambda)=$$
$$e^{s\Lambda}F^{\mathcal{O}}_{\alpha_1\alpha_2...\alpha_{n-1}\alpha_n}(\theta_1, \theta_2, \ldots \theta_{n-1}, \theta_n),$$

(5.5.4)

where s denotes the Lorentz spin of $\mathcal{O}$ [47].

- *Kinematical singularities*

  The form factors are meromorphic functions of the rapidities and have a number of kinematical poles, satisfying

$$-\frac{i}{2} \operatorname{Res}_{\theta_1=\theta_2 + i\pi} F^{\mathcal{O}}_{\bar{\alpha}_2\alpha_2...\alpha_{n-1}\alpha_n}(\theta_1, \theta_2, \ldots \theta_{n-1}, \theta_n)=$$
$$\mathbf{C}_{\bar{\alpha}_2\beta_2}\left[1 - (-)^\sigma S^{\beta_n\beta_2}_{\alpha_n\rho_{n-3}}(\theta_2 - \theta_n)...S^{\beta_3\rho_1}_{\alpha_3\alpha_2}(\theta_2 - \theta_3)\right]F^{\mathcal{O}}_{\beta_3\,\beta_4...\beta_{n-1}\,\beta_n}(\theta_3, \ldots \theta_n),$$

(5.5.5)

where $\bar{\alpha}$ indicates the antiparticle of α and $\mathbf{C}_{\bar{\alpha}\beta}$ represents the charge-conjugation matrix (in the case of interest to this book the charge conjugation will be a simple Kronecker delta). The statistical-type factor $(-)^\sigma$ is associated with graphically permuting the particle 1 through $\mathcal{O}$ (we again refer to section 7 in [45] and to the explanation in [46]). By $\mathbb{1}$ in (5.5.5), one means $S^{\beta_n\beta_2}_{\alpha_n\rho_{n-3}}...S^{\beta_3\rho_1}_{\alpha_3\alpha_2}$, with $s^{ab}_{cd} = \delta^a_c\delta^b_d$. It is quite fiddly to locate the correct indices, so the schematic contraction of S-matrices in the kinematical-singularity condition follows, for $n = 6$ as an example:

$$\mathbf{C}_{\bar{\alpha}_2\beta_2}S^{\beta_6\beta_2}_{\alpha_6\rho_3}S^{\beta_5\rho_3}_{\alpha_5\rho_2}S^{\beta_4\rho_2}_{\alpha_4\rho_1}S^{\beta_3\rho_1}_{\alpha_3\alpha_2}F_{\beta_3\beta_4\beta_5\beta_6}.$$

(5.5.6)

- *Bound state singularities*

  If the spectrum contains bound states made out of the particles present in the *in* state, then the form factors will reflect this by having additional poles. These poles are located wherever a pair of rapidities can form a bound state. The residue at these poles has to be proportional to the form factor having as the *in* state the bound state and the remaining $(n - 2)$ particle—if the first two particles form a bound state, for instance, then this means $F^{\mathcal{O}}_{\mathbf{B}, \, a_3, \, ..., \, a_n}$, where $\mathbf{B}$ is the bound state of 1 and 2. If particles 1 and 3 form a bound state, we can use the permutation formula to first bring them close to one another. For the complete description, we refer for example to [45]. This bound-state condition effectively reduces the calculation to a form factor with one less particle, the same way as the kinematical-singularity condition reduces it to two less particles. Often when such conditions are combined they are powerful enough to generate recursions that produce all the form factors starting from a few initial ones [42]. Notice that some form factors might be zero due to conservation laws or superselection rules (e.g., form factors must globally be bosonic objects).

The form-factor programme only relies on basic requisites such as the spin of the operator. This is also a slight disadvantage because one does not begin by knowing the operator and has to identify it subsequently. The spirit of the form-factor programme, very much like the *S*-matrix programme, consists in bypassing perturbation theory and simply abstracting the experience from Feynman graphs into general principles. The 'axioms' in fact derive in large part from structural requirements of local relativistic quantum field theory (for an expression of their 'raw' derivation from first postulates, the reader can once again refer to [45]). Integrability allows to further massage such postulates into a set of functional relations that we can hope to actually solve.

Correlation functions are obtained by summing over form factors, after suitably inserting resolutions of the identity expanded over multiparticle states. For instance, the two-point function is calculated as

$$\langle 0|\mathcal{O}(iR, 0)\mathcal{O}(0, 0)|0\rangle$$

$$= \sum_{n=1}^{\infty} \frac{1}{n!} \int_{-\infty}^{\infty} \frac{d\theta_1}{2\pi} \cdots \int_{-\infty}^{\infty} \frac{d\theta_n}{2\pi} \sum_{\{\alpha_i\}} F^{\mathcal{O} \, *}_{\alpha_1, \, ..., \, \alpha_n}(\theta_1, \, ... \, , \theta_n) F^{\mathcal{O}}_{\alpha_1, \, ..., \, \alpha_n}(\theta_1, \, ... \, , \theta_n) e^{-R\sum_{i=1}^{n} m_i \cos \theta_i},$$

m_i being the particle masses, and $|0\rangle$ being the true vacuum of the theory. This is thanks to the fact that in an integrable system we have control on the resolution of the identity via multiparticle states [45]:

$$\mathbb{1} = \sum_{n=1}^{\infty} \frac{1}{n!} \int_{-\infty}^{\infty} \frac{d\theta_1}{2\pi} \cdots \int_{-\infty}^{\infty} \frac{d\theta_n}{2\pi} \sum_{\{\alpha_i\}} |\theta_1, \, ... \, , \theta_n\rangle_{\{\alpha_i\}} \langle \theta_1, \, ... \, , \theta_n|_{\{\alpha_i\}}. \tag{5.5.7}$$

This programme was put to fruition, for the case that we are concerned with, in [44] and in a series of related works using the so-called *off-shell Bethe* ansatz method. We shall not discuss alternative techniques, such as the method put forward by Lukyanov [48], and Lukyanov and Zamolodchikov [13]—see also [43]. The main result of [44] is summarised as follows.

The form factor $F_{\alpha_1,\ldots,\alpha_n}(\theta_1, \ldots, \theta_n)$ for a specific *in* state is given by the component $(\alpha_1, \ldots, \alpha_n)$ of the row vector

$$N_n^{\circ} \int_{C_{\vec{\theta}}} du_1 \cdots \int_{C_{\vec{\theta}}} du_m \, g(\vec{\theta}, \underline{u}) \, \langle \Omega_n | \, C_{1\ldots n}(\vec{\theta}, u_1) \ldots C_{1\ldots n}(\vec{\theta}, u_m), \tag{5.5.8}$$

where the C operators are the C entries of the n-site monodromy matrix built with the exact R-matrix, acting from the right on the pseudovacuum *covector* $\langle \Omega_n |$— this will produce as a result a row vector instead of a column vector. The number m counts how many off-shell excitations are created over the pseudovacuum. That is to say, m counts the number of 'off-shell magnons'—off-shell waves of particles of one type moving with rapidity u_j in a sea of particles of the other type. All of the particles of the sea appear with their associated inhomogeneity/ momentum-carrying rapidity θ_i. Notice that the choice of the pseudovacuum is independent of the fact that the correlation function that one later constructs is purported to be computed on the true vacuum of the theory—the pseudovacuum being here a mere trick to solve Watson's equation. One has here employed the collective notations

$$\vec{\theta} = (\theta_1, \ldots, \theta_n), \qquad \underline{u} = (u_1, \ldots, u_m). \tag{5.5.9}$$

This method is called of the off-shell algebraic Bethe ansatz because the auxiliary parameters u_j are not fixed by the Bethe equations (that would be on-shell), but are instead integrated over. The auxiliary roots β_j (see previous section) are understood as having been solved for, their only role having been to contribute to fixing the θ_i to a true eigenstate of the spectrum.

The function g is constructed for $n > 2m$ as

$$g(\vec{\theta}, \underline{u}) = \prod_{1 \leqslant i < j \leqslant n} F(\theta_i - \theta_j) \prod_{i=1}^{n} \prod_{J=1}^{m} \phi(\theta_i - u_J) \prod_{1 \leqslant I < J \leqslant m} \tau(u_I - u_J) e^{\pm \frac{s}{n-2m}(2\sum_A u_A - \sum_k \theta_k)},$$

s being the spin of the operator, F being the minimal two-particle form-factor block,

$$\phi(u) = \frac{1}{F(u)F(u + i\pi)}, \qquad \tau(u) = \frac{1}{\phi(u)\phi(-u)}, \tag{5.5.10}$$

and the contour $C_{\vec{\theta}}$ being a complicated contour described in [44]. In the case $n \leqslant 2m$, there should be a change of the pseudovacuum (e.g., to all antisolitons) which allows this formula to be used. The minimal two-particle form-factor block satisfies the axioms for $n = 2$ with the S-matrix simply replaced by $S(\theta)$, has the minimal number of singularities and reduces to a compact integral representation [49]:

$$F(\theta) = -i \sinh \frac{\theta}{2} \times \exp \int_0^\infty \frac{dt}{t} \frac{\sinh \frac{(1-\nu)t}{2}}{\sinh \frac{\nu t}{2} \cosh \frac{t}{2}} \frac{1 - \cosh t(1 - \frac{\theta}{i\pi})}{2 \sinh t}, \qquad (5.5.11)$$

ν being defined by

$$\nu = \frac{\beta^2}{8\pi - \beta^2}. \qquad (5.5.12)$$

The minimal form-factor block $F(\theta)$ is not the minimal two-particle form factor, but it is a useful building block for it (and for arbitrary form factors in fact because it is used to systematically take care of the functional complication of the dressing phase). To obtain this formula one can utilise a theorem by Karowski and Weisz [49], which says the following. If the dressing factor can be recast in the form

$$S(\theta) = \exp \int_0^\infty dx\, f(x) \sinh \frac{x\theta}{i\pi}, \qquad (5.5.13)$$

then the minimal solution to

$$F(\theta) = F(-\theta)\, S(\theta), \qquad F(i\pi - \theta) = F(i\pi + \theta), \qquad (5.5.14)$$

is given by

$$F(\theta) = \exp \int_0^\infty dx\, f(x) \frac{\sin^2 \frac{x(i\pi - \theta)}{2\pi}}{\sinh x} \qquad (5.5.15)$$

(all formulas in this theorem being valid in appropriate regions in the complex rapidity plane).

The calculation that we are interested in is now performed by choosing an operator that is clearly identifiable. Operators that are built up as in Mandelstam's construction, which we have mentioned earlier on in this chapter, are specific enough that one can in principle pinpoint which form factor to choose and how to adapt the formula that we have just described.

On a separate ground, we can compute a tree-level Feynman graph involving the fermion field of the Thirring model, using weak-coupling perturbation theory. The paper [44] performs such a calculation (and many more were performed in the literature). At tree level, the authors find in particular

$$\langle 0|\psi(0)|\theta_1\theta_2\theta_3\rangle_{\bar{s}ss} = -ig \sinh \frac{\theta_{23}}{2} \times \frac{u(\theta_2)\cosh \frac{\theta_{12}}{2} + u(\theta_3)\cosh \frac{\theta_{13}}{2}}{\cosh \frac{\theta_{12}}{2} \cosh \frac{\theta_{13}}{2} \cosh \frac{\theta_{23}}{2}}, \qquad (5.5.16)$$

where α_i, $i = 1, 2$, denotes either a fermion or an anti-fermion, $\theta_{ij} = \theta_i - \theta_j$ and $u(\theta) = \sqrt{m_{sol}} \begin{pmatrix} e^{-\frac{\theta}{2}} \\ e^{\frac{\theta}{2}} \end{pmatrix}$ is the Dirac polarisation spinor. This expression is then compared with a strong coupling approximation argued from the exact approach—we refer to the paper [44] for a full detail of the comparison.

We are now ready to revisit the host of these observations and reported results in the conclusive chapter of our account, which will then be followed by the supplement.

References

[1] Coleman S R 1975 The quantum sine-Gordon equation as the massive Thirring model *Phys. Rev.* D **11** 2088

[2] Klassen T R and Melzer E 1993 Sine-Gordon not equal to massive Thirring, and related heresies *Int. J. Mod. Phys.* A **8** 4131–74

[3] Faber M and Ivanov A N 2001 On the equivalence between sine-Gordon model and Thirring model in the chirally broken phase of the Thirring model *Eur. Phys. J.* C **20** 723–57

[4] Faber M and Ivanov A N 2003 Is the energy density of the ground state of the sine-Gordon model unbounded from below for beta**2 greater than 8 pi? *J. Phys.* A **36** 7839

[5] Bozkaya H, Faber M, Ivanov A N and Pitschmann M 2006 On the renormalization of the two-point Green function in the sine-Gordon model *J. Phys.* A **39** 2177–201

[6] Sénéchal D 2004 An introduction to bosonization *Theoretical Methods for Strongly Correlated Electrons* (New York: Springer New York) 139–86 pp

[7] Nakanishi N 1977 Free massless scalar field in two-dimensional space-time *Prog. Theor. Phys.* **57** 269–78

[8] Weisz P H 1977 Perturbation theory check of a proposed exact Thirring model S matrix *Nucl. Phys.* B **122** 1–14

[9] Dashen R F, Hasslacher B and Neveu A 1975 The particle spectrum in model field theories from semiclassical functional integral techniques *Phys. Rev.* D **11** 3424

[10] Mandelstam S 1975 Soliton operators for the quantized sine-Gordon equation *Phys. Rev.* D **11** 3026

[11] Frohlich J and Marchetti P A 1988 Bosonization, topological solitons and fractional charges in two-dimensional quantum field theory *Commun. Math. Phys.* **116** 127

[12] Seiler R and Uhlenbrock D A 1977 On the massive Thirring model *Ann. Phys.* **105** 81

[13] Lukyanov S L and Zamolodchikov A B 2001 Form-factors of soliton creating operators in the sine-Gordon model *Nucl. Phys.* B **607** 437–55

[14] Bergknoff H and Thacker H B 1979 Method for solving the massive Thirring model *Phys. Rev. Lett.* **42** 135

[15] Bergknoff H and Thacker H B 1979 Structure and solution of the massive Thirring model *Phys. Rev.* D **19** 3666

[16] Faddeev L D 1996 How algebraic Bethe ansatz works for integrable model *Les Houches School of Physics: Astrophysical Sources of Gravitational Radiation* **5** (Les Houches: School of Physics) 149–219 pp

[17] Doikou A, Evangelisti S, Feverati G and Karaiskos N 2010 Introduction to Quantum Integrability *Int. J. Mod. Phys.* A **25** 3307–51

[18] Levkovich-Maslyuk F 2016 *The Bethe ansatz. J. Phys.* A **49** 323004

[19] Lamers J 2015 A pedagogical introduction to quantum integrability, with a view towards theoretical high-energy physics *PoS* **Modave2014:001**

[20] Miranda E 2003 Introduction to bosonization *Braz. J. Phys.* **33** 3–35

[21] Haldane F D M 1981 'Luttinger liquid theory' of one-dimensional quantum fluids. i. properties of the luttinger model and their extension to the general 1d interacting spinless fermi gas *J. Phys. C: Solid State Phys.* **14** 2585

[22] Tong D *Quantum Field Theory on a Line* https://www.damtp.cam.ac.uk/user/tong/gauge-theory/72d.pdf (accessed March 25 2024)

[23] Fradkin E 1991 Field theories of condensed matter (Cambridge: Cambridge Univ. Press)

[24] Lüscher M 1989 Bosonization in 2 + 1 dimensions *Nucl. Phys.* **B 326** 557–82

[25] Burgess C P, Lütken C A and Quevedo F 1994 Bosonization in higher dimensions *Phys. Lett.* **B 336** 18–24

[26] Gaiotto D and Kapustin A 2016 Spin tqfts and fermionic phases of matter *Int. J. Mod. Phys.* **A 31** 1645044

[27] Schultz T D, Mattis D C and Lieb E H 1964 Two-dimensional ising model as a soluble problem of many fermions *Rev. Mod. Phys.* **36** 856

[28] Lieb E, Schultz T and Mattis D 1961 Two soluble models of an antiferromagnetic chain *Ann. Phys.* **16** 407–66

[29] Fendley P and Saleur H 1993 Massless integrable quantum field theories and massless scattering in (1 + 1)-dimensions *Summer School in High-Energy Physics and Cosmology (Includes Workshop on Strings, Gravity, and Related Topics 29–30 Jul 1993)* **9** (Singapore: World Scientific) 301–32 pp

[30] Negro S 2016 Integrable structures in quantum field theory *J. Phys.* **A 49** 323006

[31] Zamolodchikov A B and Zamolodchikov A B 1992 Massless factorized scattering and sigma models with topological terms *Nucl. Phys.* **B 379** 602–23

[32] Dunning T C 2000 Perturbed conformal field theory, nonlinear integral equations and spectral problems *PhD Thesis* Durham University

[33] Tarasov V O, Takhtajan L A and Faddeev L D 1983 Local Hamiltonians for integrable quantum models on a lattice *Theor. Math. Phys.* **57** 1059–73

[34] Marboe C and Volin D 2017 Fast analytic solver of rational bethe equations *J. Phys. A: Math. Theor.* **50** 204002

[35] Li W, Okyay M and Nepomechie R I 2022 Bethe states on a quantum computer: success probability and correlation functions *J. Phys.* **A 55** 355305

[36] Ravanini F and Franzini T Thermodynamic Bethe ansatz for a family of scattering theories with $U_q(sl(2))$ symmetry *MSc thesis* (University of Bologna)

[37] Korepin V E 1980 The mass spectrum and the *S*-matrix of the massive thirring model in the repulsive case *Commun. Math. Phys.* **76** 165–76

[38] Smirnov F A 1992 *Form-Factors in Completely Integrable Models of Quantum Field Theory* vol **14** (Singapore: World Scientific)

[39] Berg B, Karowski M and Weisz P 1979 Construction of green functions from an exact *S*-matrix *Phys. Rev.* **D 19** 2477

[40] Babujian H M, Foerster A and Karowski M 2008 The form factor program: a review and new results, the nested SU(N) off-shell Bethe ansatz and the 1/N expansion *Theor. Math. Phys.* **155** 512–22

[41] Babujian H M, Foerster A and Karowski M 2008 The Nested SU(N) off-shell Bethe ansatz and exact form-factors *J. Phys.* **A 41** 275202

[42] Mussardo G 2020 *Statistical Field Theory* **3** (Oxford: Oxford University Press)

[43] Britton S and Frolov S 2013 Free field representation and form factors of the chiral Gross–Neveu model *JHEP* **11** 076

[44] Babujian H, Fring A, Karowski M and Zapletal A 1999 Exact form factors in integrable quantum field theories: the sine-Gordon model *Nucl. Phys.* B **538** 535–86

[45] Bombardelli D 2016 S-matrices and integrability *J. Phys.* A **49** 323003

[46] Doyon B 2012 Integrability. *Lecture notes for a course given at the London Taught Centre,* https://nms.kcl.ac.uk/benjamin.doyon/ltcc/notes-integrability.pdf (accessed on 3 March 2023)

[47] Dorey P 1996 Exact S matrices *Eotvos Summer School in Physics: Conformal Field Theories and Integrable Models* (Berlin: Springer) 8 85–125 pp

[48] Lukyanov S 1995 Free field representation for massive integrable models *Commun. Math. Phys.* **167** 183–226

[49] Karowski M and Weisz P 1978 Exact form-factors in (1 + 1)-dimensional field theoretic models with soliton behavior *Nucl. Phys.* B **139** 455–76

IOP Publishing

Integrability using the Sine-Gordon and Thirring Duality
An introductory course
Alessandro Torrielli

Chapter 6

Remarks on the duality

The duality between the Sine-Gordon and the Thirring model is a remarkable chapter of theoretical physics. In this book we have tried to portray it in hopefully accessible terms, only touching on some of the subtleties, which the reader is encouraged to independently explore. Some of these subtleties are delved into in great detail for instance in [1], see also the end of section 1.3 in [2]. We also again refer to [3], especially page 369 and following. Technically, what we have presented should be understood as a duality or a correspondence, and not as an equivalence. The two theories are not related by a (local) change of variables because the 'dictionary' that we have discussed does not qualify as such. According to a stricter interpretation of 'duality' or 'correspondence', the one between Sine-Gordon and Thirring also probably does not qualify as such. Rather, certain observables in the zero-charge sectors of either theory obey precise matching patterns. But there are differences in other aspects [1, 2]. We hope that this account will be a sufficient starting point for the reader's individual examination of the field. Our pedagogical choice has been to highlight the commonalities, with the primary purpose of instilling the idea of duality by highlighting some of the features of its ground-breaking impact, and the realisation of the power of integrability in analysing such statements.

6.1 The paper by Klassen and Melzer

The paper [1] took the task of re-examining the previous literature under the suggestive title of 'Sine-Gordon $\neq$ massive Thirring and related heresies'. This paper clarifies the point that we have made above that the two theories are not in fact exactly equivalent. Let us summarise the salient points of Klassen and Melzer's analysis.

The starting observation is to point out precisely what Coleman and Mandelstam, respectively, *did* prove. On the one hand, Coleman's work shows that the correlation functions of the fermion bilinear $\bar{\psi}_T \psi_T$ in the massive Thirring model, and, respectively, of the field $\cos \beta \phi$ in Sine-Gordon, when appropriately dressed as in the second relation in (5.1.2), are the same in their, respective, conformal perturbation theories. On the other hand, Mandelstam's work shows that one can construct anticommuting operators, satisfying the same relations as the fundamental fermion field of the Thirring model, by using the field φ of the massive Sine-Gordon theory.

This is not sufficient to conclude that the two theories are equivalent, and in fact they are not. Klassen and Melzer proceed to observe that the ultraviolet limit of the Sine-Gordon model and of Thirring are two *distinct* $c = 1$ conformal field theories (CFTs): in the case of Sine-Gordon one obtains the *bosonic* Gaussian model, while in the case of Thirring one obtains the *fermionic* Gaussian model[1]. Nonetheless, these two distinct models have a common subsector whose vertex operators can be identified. This is the case for instance of $V_{0,1} + V_{0,-1} \sim e^{i\beta\phi} + e^{-i\beta\phi} \sim \cos \beta\phi$, which can be located in both CFTs. In [1]—see equations (3.3) and (3.4) in that paper—the precise way in which adding this perturbation precisely corresponds to perturbing the massless Thirring model with a $\bar{\psi}_T\psi_T$ term is elucidated. Perturbing the two distinct bosonic and fermionic CFTs in this fashion gives the two distinct massive models: Sine-Gordon and Thirring, respectively. All the correlation functions of the common perturbing operator are, however, the same in the two models, leading to two identical conformal perturbation theories to all orders (this also applies separately to $V_{0,1}$ and $V_{0,-1}$). Coleman's result is directly connected with this statement.

When it comes to Mandelstam's result, Klassen and Melzer make a point that there are two distinct types of vertex operators in the two respective CFTs which, despite all having ± 1 values of the $\mathfrak{u}(1)$ (topological) charge, are not common to the two CFTs.

1. One (first) type of such vertex operators is given by $V_{\pm 1,0}$ and is present in the bosonic CFT: their massive version is proposed by [1] as creating the soliton and the antisoliton, repectively.

2. Another (second) type of such vertex operators is given by $V_{\pm 1,\frac{1}{2}}$ and $V_{\pm 1,-\frac{1}{2}}$ and is present in the fermionic CFT: their massive version will create the Thirring fermions (in two possible inequivalent ways, see formula (2.16) in [1] combined with the second paragraph of page 12 in that paper). In fact, Klassen and Melzer state that the bosonised[2] components of the massive spin $\frac{1}{2}$ Dirac field creating the fermionic (anticommuting) particles are the massive versions of the vertex operators of this second type.

Notice that the statistics of these vertex operators is

$$V_{m,n}(x) V_{p,q}(y) = (-)^{2mq} \, V_{p,q}(y) V_{m,n}(x), \qquad x \neq y, \qquad (6.1.1)$$

(once appropriate Klein factors are taken into account, as per the discussion above formula (2.7) in [1]) and therefore $V_{\pm 1,0}$ commute while $V_{\pm 1,\frac{1}{2}}, V_{\pm 1,-\frac{1}{2}}$ all anticommute. The spin of the operators is in mn. The two types of vertex operators are not *mutually local*, which is a statement about the (non-)single-valuedness of the OPE

---

[1] We also refer the reader to [4, 5], where this relationship is recast in terms of perturbations of the $\mathfrak{su}(2)$ Wess–Zumino–Witten (WZW) model.

[2] In the context of vertex operators in CFT, the terminology 'bosonised' quite literally indicates 'being written in terms of a bosonic field', as in Mandelstam's construction.

coefficients. In this case the constraint of mutual locality (to have single-valuedness) between V_{mn} and V_{pq} is $mq + np \in \mathbb{Z}$, which is violated across two different types. It is not implausible that either type can, however, be expressed in terms of the same building block φ. Mandelstam's result is an incarnation of this statement. Of course, as pointed out by [6], we have surreptitiously passed the confines of applicability of the spin-statistics theorem the moment we started dealing with nonlocal operators. We have implicitly here assumed a sort of continuity of the notion of spin-statistics with the CFT version, based on the properties displayed by the vertex operators $V_{m,n}$.

Therefore, we see that physically the picture is clear and it connects with what we have been discussing earlier in this book. The essential fermionic behaviour of the solitons is determined by their being created by bosonic (commuting) vertex-type operators built out of φ, but having interactions that determine an $S(0) = -1$ scattering matrix. Mandelstam showed that one can construct anticommuting operators that are also built out of φ, which are bosonised (see footnote 2) versions of the (spin $\pm\frac{1}{2}$ anticommuting) Thirring fermionic field components. It is to be remarked that the technique presented in Mandelstam's paper can easily be employed to produce massive versions of $V_{\pm 1,0}$, also expressed in terms of φ. For example, the operator

$$\exp\left[-\frac{2\pi i}{\beta} \int_{-\infty}^{x} d\xi\, \partial_t \phi(\xi,\, t) + i\beta\phi(x,\, t) \right] \tag{6.1.2}$$

(always undertood as normal-ordered and appropriately regulated) commutes with itself at a different spatial location [7]. Along these lines, it is therefore clear that the S-matrix for the scattering of fermions in the Thirring model should have the opposite sign as the Sine-Gordon S-matrix—and an otherwise equal functional form. The scattering involving bound states does not present any sign issue and is identical in the two theories.

These considerations on the intrinsic differencies between Sine-Gordon and Thirring of course extend to the 'free-fermion' point $\beta^2 = 4\pi$. There in fact the Sine-Gordon solitons are bosons with an S-matrix identically equal to -1 (hence they are bosons that behave like free fermions).

At this stage, we should return to a point that we had left open when describing Mandelstam's construction as regards to the periodicity of the field φ and the well-definiteness of the exponential operators. We can start by considering the CFT vertex operators, which should be the CFT analogue of Mandelstam's operators, and in particular we can focus on $V_{1,\frac{1}{2}}$ for example. The massive version of this vertex operator will correspond to a Thirring fermion. In the language and the precise notation of Klassen and Melzer [1], such an operator is given by

$$V_{1,\frac{1}{2}} = \,: e^{i\frac{\Phi}{r} + ir\tilde{\Phi}} :. \tag{6.1.3}$$

In terms of the Sine-Gordon field, one has $\phi = \frac{\tilde{\Phi}}{\pi}$. The crucial point is that the definition of the fields Φ and $\tilde{\Phi}$ is such that

$$(\Phi,\, \tilde{\Phi}) \sim (\Phi,\, \tilde{\Phi}) + \left(2\pi nr,\, \frac{\pi m}{r} \right), \tag{6.1.4}$$

where m, n belong to a particular lattice: either $m \in 2\mathbb{Z}$ and $n \in \mathbb{Z}$, or $m \in 2\mathbb{Z} + 1$ and $n \in \mathbb{Z} + \frac{1}{2}$. We can easily see that for every point of this lattice the vertex operator (6.1.3) is well-defined, meaning that it acquires the same value. There are sometimes minuses that come from one shift, but they are always compensated by the other shift. For example,

$$(\Phi, \tilde{\Phi}) \sim (\Phi, \tilde{\Phi}) + \left(3\pi r, \frac{3\pi}{r}\right) \tag{6.1.5}$$

corresponds to

$$V_{1,\frac{1}{2}} \sim V_{1,\frac{1}{2}} \times e^{3i\pi} \times e^{3i\pi} \sim V_{1,\frac{1}{2}}, \tag{6.1.6}$$

and so all is good. The important lesson is that the field Φ has to do its part in the periodicity transformation to compensate. Therefore, we would have been naive if we had ever thought that the integral part of the exponent in Mandelstam's construction would not bear some periodicity. The more careful consideration is that the conjugate momentum $\pi_\phi = \dot{\phi}$ must also be subject to a suitable periodicity condition for the consistency of the operator algebra.

In the case where breathers are considered, we can also return to a statement that we made as regards to the quantum of the field φ. In view of the UV CFT perspective, it is appropriate to reinterpret the bound states in terms of CFT vertex operators. This is dealt with in detail in section 4.4 of [1].

We conclude this part by returning to another point which we have left open, concerning the spectrum of quantum Sine-Gordon. We promised that we would come back to this issue and rephrase in a CFT language. This is now possible, thanks to the vertex operators $V_{m,n}$. Here we also use [8].

In the $c = 1$ CFT corresponding to the free compact boson the operators $V_{m,n}$ label the momentum and winding quantum numbers of the modes of the field—see also [9]. To each pair (m, n) is associated a different superselection sector of the Fock space, and the total Fock space is the direct sum of all the superselection sectors. The vertex operators create the ground state of each sector from the vacuum state $|0\rangle$:

$$|m, n\rangle = V_{m,n}|0\rangle. \tag{6.1.7}$$

Within each superselection sector, the modes of the field create excitations over the (n, m) th ground state.

As we have just seen the Sine-Gordon model and the Thirring model correspond to two different $c = 1$ CFTs in the UV, with vertex operators corresponding to two distinct subsets of the $V_{m,n}$ s. If we focus on the Sine-Gordon model, Mandelstam has constructed massive analogues of the vertex operators $V_{\pm 1,0}$ that create a single soliton and a single antisoliton, respectively. The Sine-Gordon is a deformation of the free boson, and the deformation admits the conservation of the topological charge, which counts the number of solitons and antisolitons. The topological charge is the analogue of the winding number. We therefore see that the massive analogues of the ground states in each superselection sector are the lowest energy configurations in which a fixed number of solitons and antisolitons can be found. This is the analogue of the classical solitons interpolating between minima of the potential: the multisoliton states and their excitations populate each superselection sectors.

The momentum quantum number is associated with the charge $\pi_0 = \int_{-\infty}^{\infty} dx \, \partial_0 \phi$, and it is not conserved in the massive theory. The Noether symmetry $\phi \to \phi + constant$ is broken by the cosine term for arbitrary constant shifts and is only preserved for a discrete subset of shifts compatible with the cosine periodicity— call these shifts $s_{n \in \mathbb{Z}} \leftrightarrow \frac{2\pi n}{\beta}$. There will therefore be no Noether charge in the massive theory. Nevertheless, it is still fruitful to use the vertex operators and their CFT labels in the framework of conformal perturbation theory, where at every order one computes correlators in the CFT.

An additional effect needs to be included to complete the full description of the Fock space, which can be found discussed in section 4 of [10] for instance. This effect is related to the way in which the symmetry $s_{n \in \mathbb{Z}}$ is realised on the quantum states. The complete Fock space is further organised according to a *quasimomentum* or *twist* $-\pi < \alpha \leqslant \pi$. Each quasimomentum sector has a lowest-energy state (α-ground state or α-vacuum). In infinite volume, each quasimomentum sector is identical and degenerate to any another, each consisting of a vacuum and multi-solitons states. The distinction arises for finite volume, as explained in [10], see also [11–13], and a nice discussion in section 4.2 of [14]. In [13], the authors discussed how the degeneracy amongst the quasimomentum sectors is lifted by putting the theory on a finite circle L_0—see also [15]. The true ground state of the theory is then found in the $\alpha = 0$ sector. The UV limit of the effective central charge computed from the TBA (or the Destri–de Vega integral equation, as done in [15]) shows the limit $c \to 1 - \frac{6\xi}{(\xi + \pi)} \frac{\alpha^2}{\pi^2}$ when extracted from the α-sector ground state energy. This consistently equals the ground state contribution plus (remember that the central charge is extracted from minus the TBA energy) the contribution from the dimension of the operators in the UV CFT which create the α-vacuum from the true vacuum.

6.2 Final remarks

We wish to conclude by saying that we have not made any attempt at completeness when citing the literature, which would be an impossible task. This topic was and still is, even after so many years of the community getting to grips with dualities (primarily the AdS/CFT duality [16–18]), a striking demostration of the versatility of quantum field theory, which could indeed be considered a precursor of many modern developments. To signify the subversive nature of this idea, it is worth repeating one sentence from Coleman's acknowledgements in [6]:

> *'Jeffrey Goldstone, Roman Jackiw, and Andre Neveu for discussion […],*
> *Howard Georgi for guiding me […], Konrad Osterwalder for reassuring*
> *me of my sanity.' (from [6]).*

The thought of these giants reassuring each other of their sanity is certainly one of the most impressive images of the impact that this discovery has had.

References

[1] Klassen T R and Melzer E 1993 Sine-Gordon not equal to massive Thirring, and related heresies *Int. J. Mod. Phys.* A **8** 4131–74

[2] Feverati G 2000 *Finite Volume Spectrum of Sine-Gordon Model and Its Restrictions* PhD thesis (University of Bologna)

[3] Tong D *Quantum Field Theory on a Line* https://www.damtp.cam.ac.uk/user/tong/gauge-theory/72d.pdf (accessed March 25 2024)

[4] Kobayashi K-i and Uematsu T 1990 Higher integrals of motion in a perturbed k = 1 SU(2) Wess-Zumino-Witten theory *Mod. Phys. Lett.* A **5** 823

[5] Kobayashi K-i and Uematsu T 1990 Conservation laws in a perturbed $k = 1$ SU(2) Wess–Zumino–Witten model *KEK Workshop on Topology, Field Theory and Superstring Theory* (Tsukuba: National Laboratory for High Energy Physics)

[6] Coleman S R 1975 The quantum sine-Gordon equation as the massive Thirring model *Phys. Rev.* D **11** 2088

[7] Mandelstam S 1975 Soliton operators for the quantized sine-Gordon equation *Phys. Rev.* D **11** 3026

[8] Feverati G, Ravanini F and Takacs G 1999 Nonlinear integral equation and finite volume spectrum of sine-Gordon theory *Nucl. Phys.* B **540** 543–86

[9] Sénéchal D 2004 An introduction to bosonization *Theoretical Methods for Strongly Correlated Electrons* (New York: Springer) pp 139–86

[10] Zamolodchikov A B 1994 Painlevé III and 2D polymers *Nucl. Phys.* B **432** 427–56

[11] Bazhanov V V, Lukyanov S L and Zamolodchikov A B 1996 Integrable structure of conformal field theory, quantum KdV theory and thermodynamic Bethe ansatz *Commun. Math. Phys.* **177** 381–98

[12] Bazhanov V V, Lukyanov S L and Zamolodchikov A B 1997 Integrable quantum field theories in finite volume: excited state energies *Nucl. Phys.* B **489** 487–531

[13] Fioravanti D, Mariottini A, Quattrini E and Ravanini F 1997 Excited state Destri-De Vega equation for sine-Gordon and restricted sine-Gordon models *Phys. Lett.* B **390** 243–51

[14] Conti R, Negro S and Tateo R 2019 Conserved currents and $T\bar{T}_s$ irrelevant deformations of 2D integrable field theories *JHEP* **11** 120

[15] Ravanini F 2001 Finite size effects in integrable quantum field theories *Non-perturbative QFT Methods and Their Applications* (Singapore: World Scientific) pp 199–264

[16] Maldacena J M 1998 The large N limit of superconformal field theories and supergravity *Adv. Theor. Math. Phys.* **2** 231–52

[17] Aharony O, Gubser S S, Maldacena J M, Ooguri H and Oz Y 2000 Large *N* field theories, string theory and gravity *Phys. Rep.* **323** 183–386

[18] D'Hoker E and Freedman D Z 2002 Supersymmetric gauge theories and the AdS/CFT correspondence *Theoretical Advanced Study Institute in Elementary ParticlePhysics (TASI 2001): Strings, Branes and EXTRA Dimensions* (Singapore: World Scientific) pp 3–158

IOP Publishing

Integrability using the Sine-Gordon and Thirring Duality
An introductory course
Alessandro Torrielli

Chapter 7

Supplement: the residue of the Lee–Yang model

7.1 Pole analysis

In this part of the supplement we discuss the residue at the pole of the Lee–Yang S-matrix and its sign.

Let us start by expressing the S-matrix (2.2.2) of the Lee–Yang model in terms of the Mandelstam variable s:

$$S_{LY}(s) = \frac{-\sqrt{3}\,m^2 - s\sqrt{-1 + \dfrac{4m^2}{s}}}{\sqrt{3}\,m^2 - s\sqrt{-1 + \dfrac{4m^2}{s}}} = \frac{s^2 - 4sm^2 - 3m^4 - 2\sqrt{3}\,sm^2\sqrt{-1 + \dfrac{4m^2}{s}}}{(s - m^2)(s - 3m^2)},$$

$$s = 2m^2(1 + \cosh\theta).$$

In this formula we have initially chosen $s \in (0, 4m^2)$ and then continued the expression to the whole physical region away from branch cuts. We can see that there are two poles, one in the s-channel at $s = m^2 = s_0$ and one in the t-channel at $s = 3m^2 = 4m^2 - s_0$. We can also see that there are two cuts, precisely as we expect from general principles: the s-channel cut starts at $s = 4m^2$ and goes to infinity along the positive real axis, while the t-channel cut starts at $s = 0$ and goes to infinity along the negative real axis—see figure 7.1.

The residue at the simple pole $s = m^2$ is given by $+6m^2$. According to [1], we expect the residue to be governed by

$$S_{LY} \sim -\frac{m^4}{2m_0\sqrt{4m^2 - m_0^2}}\frac{g_0^2}{s - m_0^2}, \tag{7.1.1}$$

where in this case $m_0 = m$. Therefore, we expect

$$S_{LY} \sim -\frac{m^2}{2\sqrt{3}}\frac{g_0^2}{s - m^2}. \tag{7.1.2}$$

doi:10.1088/978-0-7503-5899-6ch7

```
In[3]:= s = a + I b;
        m = 1;
```

$$\text{Plot3D}\left[\text{Re}\left[\frac{-\sqrt{3}\ m^2 - \sqrt{-1 + \frac{4 m^2}{s}}\ s}{\sqrt{3}\ m^2 - \sqrt{-1 + \frac{4 m^2}{s}}\ s}\right],\ \{a, -8, 8\},\ \{b, -5, 5\}\right]$$

$$\text{Plot3D}\left[\text{Im}\left[\frac{-\sqrt{3}\ m^2 - \sqrt{-1 + \frac{4 m^2}{s}}\ s}{\sqrt{3}\ m^2 - \sqrt{-1 + \frac{4 m^2}{s}}\ s}\right],\ \{a, -8, 8\},\ \{b, -5, 5\}\right]$$

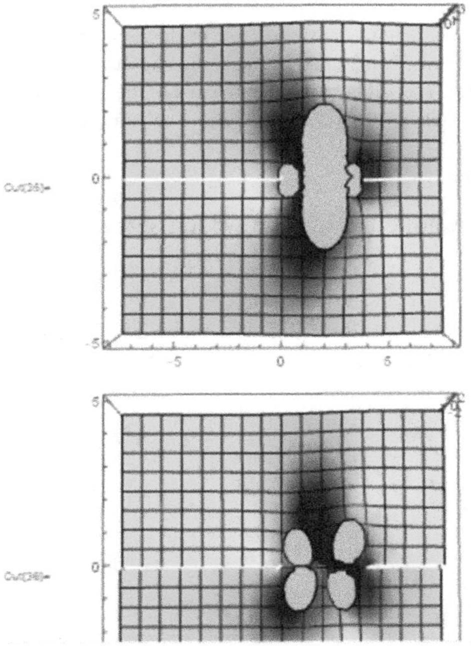

Figure 7.1. The cut structure of the Lee–Yang S-matrix in the Mandelstam s plane is shown for a convenient choice of the units of measurement $m = 1$.

In particular, we expect the residue to be negative if the three-point coupling g_0 of the theory corresponding to the process

$$\text{LY} + \text{LY} \longrightarrow \text{LY}, \qquad \text{strength}\ g_0 \tag{7.1.3}$$

(where LY is the Lee–Yang particle) is meant to be real—see figure 7.2. However, we see that the residue that we have calculated is positive, which means that

$$-g_0^2 \frac{m^2}{2\sqrt{3}} = 6m^2. \tag{7.1.4}$$

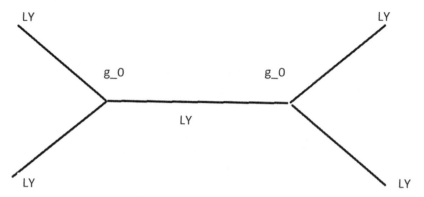

Figure 7.2. The virtual exchange corresponding to the simple pole of the S-matrix at $s = m^2$. This tree-level diagram is an effective description: the parameters g_0 and the mass m are exact (nonperturbative) because we are using the exact S-matrix of the theory. The virtual particle in the case of the Lee–Yang model is the same as the external particles. The three-point coupling is g_0, which turns out to be purely imaginary for the Lee–Yang model.

This is what is meant by the residue having the wrong sign.

Reference

[1] Paulos M F, Penedones J, Toledo J, van Rees B C and Vieira P 2017 The S-matrix bootstrap II: two dimensional amplitudes *JHEP* **11** 143

IOP Publishing

Integrability using the Sine-Gordon and Thirring Duality
An introductory course
Alessandro Torrielli

Chapter 8

Supplement: Hopf algebra properties

8.1 Building blocks

We shall try here to distill some of the necessary building blocks of quantum group theory. We will not aim at mathematical rigour but will select those parts that are relevant for the physical applications to scattering theory and to the Bethe ansatz. We will mostly confine ourselves to bosonic algebras.

Excellent references on Hopf algebras are provided, for instance, by [1–8].

Let us start by reminding ourselves that there is a mathematical way of formulating the operation of multiplying together Lie-algebra generators. Normally, this is not an allowed operation because one is only allowed to take Lie brackets of Lie-algebra elements. The enveloping algebra $U(\mathfrak{g})$ of a Lie algebra $\mathfrak{g}$ is intuitively the space of all the products of Lie-algebra elements, but mathematically it has a proper definition that makes it clearer. One starts with the so-called *tensor algebra* $T(\mathfrak{g})$ associated with a Lie algebra $\mathfrak{g}$. The tensor algebra contains all the elements of the type

$$a_1 \otimes b_1 \otimes c_1 \otimes \cdots + a_2 \otimes b_2 \otimes c_2 \otimes \cdots + \cdots, \qquad (8.1.1)$$

with an arbitrary finite number of factors in each tensor product and an arbitrary finite number of terms in the sum. Each of the a_i, b_i, c_i, ... belong to $\mathfrak{g}$. For instance,

$$a \otimes b + c \quad \text{and} \quad t \otimes u \otimes v + g \otimes g \otimes f \otimes g \qquad (8.1.2)$$

are two randomly chosen elements of the tensor algebra of $\mathfrak{g}$, provided all the elements which appear in (8.1.2), namely, a, b, c, t, u, v, g, f, belong to $\mathfrak{g}$. To obtain the universal enveloping algebra $U(\mathfrak{g})$ we mod out $T(\mathfrak{g})$ by a two-sided ideal. This ideal $\mathcal{I}$ is defined by

$$T(\mathfrak{g}) \otimes \left(T^a \otimes T^b - T^b \otimes T^a - f^{ab}_c T^c\right) \otimes T(\mathfrak{g}), \qquad a, b = 1, \ldots, dim[\mathfrak{g}], \qquad (8.1.3)$$

where T^a are the generators of $\mathfrak{g}$, f^{ab}_c the structure constants, and $dim[\mathfrak{g}]$ is the dimension of $\mathfrak{g}$ (the Einstein summation convention is applied). If we look at this

doi:10.1088/978-0-7503-5899-6ch8

carefully, we understand what this definition is doing. The tensor algebra simulates our intuition of multiplying together elements of a Lie algebra, but by itself simply juxtaposing elements does not quite work. To really simulate the multiplication of Lie-algebra elements we need to enforce that each combination within the brackets in (8.1.3) vanishes. We impose this by forcing it to be in the ideal (hence, to effectively be set to zero), everywhere it appears in any string (or word) in $T(\mathfrak{g})$. Therefore, since it vanishes, it sets the Lie bracket—which is equal to $f_c^{ab} T^c$—also equal to the result obtained by juxtaposing two elements in two different ways and subtracting. We have in this way created a *product* that we can use to compute the Lie bracket as a commutator. Practically, modding out by (9.4.5) sets $[a, b]$ equal to $ab - ba$. It is standard to denote the product or multiplication, such as the one that we have just constructed, as $\mu(a, b) = ab$.

The universal enveloping algebra of a Lie algebra is not a Lie group, even though it has some features in common with it. For instance, the process of exponentiation occuring in going from a Lie algebra to the associated Lie group also intuitively involves taking products (powers) of Lie algebra generators, when the exponential is expanded in Taylor series. However, despite this superficial similarity, the two structures remain quite distinct. Nevertheless, by equipping the universal enveloping algebra with a suitable notion of topology, one can incorporate infinite series, and therefore exponentials. Such exponentials do tend to have *group-like* features; for instance, they can be endowed with a very particular coproduct. A *group-like* coproduct is such that

$$\Delta(g) = g \otimes g. \tag{8.1.4}$$

The identity element 1 belongs to $U(\mathfrak{g})$ and is one such group-like element (as the identity would naturally be). Moreover, roughly, if $g = \exp a$, then it is not difficult to show, by the property of homomorphism of the coproduct, that

$$\Delta(a) = a \otimes 1 + 1 \otimes a, \tag{8.1.5}$$

which is the standard coproduct of a Lie algebra generator (akin to the Leibniz rule of derivatives—derivatives are indeed infinitesimal displacements in the tangent space directions). In fact

$$\Delta(g) = \exp \Delta(a) = \exp [a \otimes 1 + 1 \otimes a] = \exp [a \otimes 1] \exp [1 \otimes a] = g \otimes 1. 1 \otimes g = g \otimes g,$$

where in one of the intermediate stages we have used the fact that $a \otimes 1$ and $1 \otimes a$ commute—because they would act nontrivially on different spaces of a tensor-product representation. The coproducts (8.1.4) and (8.1.5) are characteristic of a Lie-group symmetry action, and, respectively, a Lie-algebra symmetry action in ordinary quantum mechanics.

For our purposes, coproducts will always be required to be *co-associative*, namely

$$(\Delta \otimes 1)\Delta = (1 \otimes \Delta)\Delta, \tag{8.1.6}$$

where the composition of maps is understood. We often denote by 1 the identity in the appropriate space (this could be as a map or as an element of $U(\mathfrak{g})$ for instance), except for the identity in $\mathbb{C}$—the field of coefficients for all our scopes—where the identity is of course denoted by 1. It is clear that (8.1.5) satisfies the coassociativity requirement:

$$(\Delta \otimes \mathbb{1})\Delta(T^a) = T^a \otimes \mathbb{1} \otimes \mathbb{1} + \mathbb{1} \otimes T^a \otimes \mathbb{1} + \mathbb{1} \otimes \mathbb{1} \otimes T^a = (\mathbb{1} \otimes \Delta)\Delta(T^a). \quad (8.1.7)$$

This property is the analogue of what associativity is for a multiplication map $\mu(a \otimes b) = ab$, i.e., $\mu(\mu \otimes \mathbb{1})(a \otimes b \otimes c) = \mu(\mathbb{1} \otimes \mu)(a \otimes b \otimes c)$ or simply $(ab)c = a(bc)$. We will not explore here structures that do not respect coassociativity, cf [9–11].

Notice that the coproducts (8.1.4) and (8.1.5) are cocommutative because they trivially satisfy $\Delta^{op} = \Delta$. The multiplication is, however, not commutative because $T^b T^a \neq T^a T^b$ in principle—we could say that $\mu^{op}(T^a \otimes T^b) \neq \mu(T^a \otimes T^b)$, where $\mu^{op} = \mu \circ \Pi$, Π being the permutation of the two factors of the tensor product. We have already encountered earlier coproducts that are not cocommutative. In that case, it remains true that the opposite coproduct does define another perfectly valid coproduct map, although distinct from the original one.

If the coproduct represents the action of the symmetry generators on a two-particle state, then there are other maps that define a Hopf algebra but which may or may not have a direct physical interpretation. Some are necessary, primarily for mathematical consistence.

An *algebra* is defined by the multiplication μ and the unit η. As we have seen, $U(\mathfrak{g})$ is an algebra because it has been endowed with a notion of multiplication:

$$\mu: U(\mathfrak{g}) \otimes U(\mathfrak{g}) \rightarrow U(\mathfrak{g}). \quad (8.1.8)$$

We just write

$$\mu(a \otimes b) \equiv ab \quad \text{or} \quad a.b \quad (8.1.9)$$

as a shorthand notation. Notice that we will also write AB in the case of two maps A and B, to mean the composition of such maps. The context should always clarify which multiplication one means. $U(\mathfrak{g})$ also admits a unit

$$\eta: \mathbb{C} \rightarrow U(\mathfrak{g}), \quad (8.1.10)$$

such that $\eta(1) = \mathbb{1}$.

A *coalgebra* is defined by a comultiplication (coproduct) and a counit, for which effectively adding the prefix *co-* reverses the arrows:

$$\Delta: U(\mathfrak{g}) \rightarrow U(\mathfrak{g}) \otimes U(\mathfrak{g}), \qquad \varepsilon: U(\mathfrak{g}) \rightarrow \mathbb{C}. \quad (8.1.11)$$

The important relation of compatibility that characterises a coalgebra is given by

$$(\varepsilon \otimes \mathbb{1})\Delta = (\mathbb{1} \otimes \varepsilon)\Delta = \mathbb{1}. \quad (8.1.12)$$

It is easy to convince oneself that the counit that is appropriate to $U(\mathfrak{g})$ is given by

$$\varepsilon(\mathbb{1}) = 1, \qquad \varepsilon(T^a) = 0, \quad (8.1.13)$$

and then extended by homomorphism to the whole of $U(\mathfrak{g})$—because we need to have that $\varepsilon(ab) = \varepsilon(a)\varepsilon(b)$. It is easy to prove (8.1.13):

$$(\varepsilon \otimes \mathbb{1})\mathbb{1} \otimes \mathbb{1} = 1 \otimes \mathbb{1} = \mathbb{1}, \quad (8.1.14)$$

and

$$(\varepsilon \otimes 1)(a \otimes 1 + 1 \otimes a) = 0 \otimes 1 + 1 \otimes a = a, \qquad (8.1.15)$$

where we have used the fact that tensoring an element of $\mathbb{C}$ with an element of $U(\mathfrak{g})$ just amounts to the ordinary multiplication by a coefficient. If a space is both an algebra and a coalgebra, then it is automatically a *bialgebra*.

To turn a bialgebra into a Hopf algebra, one needs the antipode map Σ. The antipode is uniquely defined by the coproduct because it has to satisfy the relation

$$\mu(\Sigma \otimes 1)\Delta = \eta\varepsilon. \qquad (8.1.16)$$

We can, as an exercise, simply apply this relation to $U(\mathfrak{g})$ and find

$$\mu(\Sigma \otimes 1)\Delta(1) = \eta\varepsilon(1) \qquad \text{hence} \qquad \Sigma(1)1 = 1 \qquad \text{hence} \qquad \Sigma(1) = 1,$$

and

$$\mu(\Sigma \otimes 1)\Delta(T^a) = \eta\varepsilon(T^a) \qquad \text{hence} \qquad \Sigma(T^a)1 + \Sigma(1)T^a = 0$$
$$\text{hence} \qquad \Sigma(T^a) = -T^a.$$

This clarifies the idea that the antipode is mathematically related to taking the inverse of a group-like element because it preserves the identity and changes the sign to a Lie-algebra generator. The physical connection with antiparticles comes from noticing that, given a certain (*particle*) representation of $U(\mathfrak{g})$, let us call it ρ such that $\rho(T^a)$ is a given matrix, the representation of the antipode of T^a is a matrix of the same dimension. Therefore, we can associate with the antipode an *antiparticle* representation $\bar{\rho}$, via the rule:

$$\rho(\Sigma(T^a)) = C^{-1}[\bar{\rho}(T^a)]^{tr}C, \qquad (8.1.17)$$

where C is the *charge conjugation matrix* and ^{tr} denotes transposition. Transposition and similarity transformations also preserve a matrix dimension, and the formula (8.1.17) inter-relates all these matrices of the same dimension. The physical situation determines both the charge conjugation matrix and the representation $\bar{\rho}$ such that (8.1.17) is satisfied.

It can be proven that the antipode is an anti-homomorphim:

$$\Sigma(ab) = \Sigma(b)\Sigma(a) \qquad (8.1.18)$$

(always working in the nongraded setting).

The last comment that we make here is that the same set of maps and relations as we have been outlining is used to define any other Hopf algebra. We have described here the basic example of $U(\mathfrak{g})$, but deformations thereof, and more general spaces, provided that they are equipped with admissible such maps which satisfy the given relations, will qualify as Hopf algebras. We have seen a deformation which goes under the name of $U_q(\mathfrak{sl}(2))$ in the case of Sine-Gordon or the XXZ spin-chain—see section 3.6, but there are more general cases that encompass a great variety of integrable systems.

8.2 Coproducts

In this part of the supplement we collect some more properties of Hopf algebras.

We begin with the property of the standard coproduct (8.1.5) to be a Lie algebra homomorphism. This is easy to show in full generality:

$$
\begin{aligned}
[\Delta(T^a), \Delta(T^b)] &= [T^a \otimes 1 + 1 \otimes T^a, T^b \otimes 1 + 1 \otimes T^b] \\
&= [T^a, T^b] \otimes 1 + 1 \otimes [T^a, T^b] \\
&= i f_c^{ab}(T^c \otimes 1 + 1 \otimes T^c) = i f_c^{ab} \Delta(T^c),
\end{aligned}
\tag{8.2.1}
$$

showing that the coproduct provides a representation of the Lie algebra. We have used the induced multiplication on the tensor product:

$$
a \otimes b . c \otimes d = ac \otimes bd
\tag{8.2.2}
$$

(no fermionic signs are present in our formulas because we are not dealing with Lie superalgebras at this stage).

The statement (8.2.1) is nothing else than the realisation that tensoring two representations gives a third (possibly reducible) representation of the same Lie algebra. In fact, we can choose two different matrix representations ρ_1 and ρ_2, giving rise to matrices of dimensions d_1 and d_2, respectively, for the two factors of the tensor product. Hence, from $\rho_1(T^a)$ and $\rho_2(T^a)$ we obtain a third representation

$$
\rho_\otimes(T^a) = \rho_1(T^a) \otimes 1_{d_2} + 1_{d_1} \otimes \rho_2(T^a),
\tag{8.2.3}
$$

the indices explicitly denoting the dimensions of the identity matrices.

When we deform the standard coproduct to obtain a more general quantum group, things become a bit more complicated. Let us prove the homomorphism property for the relation $[H, E] = E$, now seen as one of the relations (3.6.1) defining $U_q(\mathfrak{su}(2))$. The coproduct that we now use is the deformed one that we have first encountered, so we have

$$
\begin{aligned}
[\Delta(H), \Delta(E)] &= [H \otimes 1 + 1 \otimes H, E \otimes q^{-H} + q^H \otimes E] \\
&= [H, E] \otimes q^{-H} + q^H \otimes [H, E] \\
&= E \otimes q^{-H} + q^H \otimes E = \Delta(E),
\end{aligned}
\tag{8.2.4}
$$

proving the homomorphism property. We have exploited the fact that H clearly commutes with $q^{\pm H}$. The remaining relations proceed analogously, with a slight further complication when it comes to $[E, F]$:

$$
\begin{aligned}
[\Delta(E), \Delta(F)] &= [E \otimes q^{-H} + q^H \otimes E, F \otimes q^{-H} + q^H \otimes F] \\
&= [E, F] \otimes q^{-2H} + q^H F \otimes E q^{-H} - F q^H \otimes q^{-H} E \\
&\quad + E q^H \otimes q^{-H} F - q^H E \otimes F q^{-H} + q^{2H} \otimes [E, F].
\end{aligned}
\tag{8.2.5}
$$

We now use

$$
q^{\pm H} E = \sum_{n=0}^{\infty} \frac{(\pm H)^n}{n!} E = \sum_{n=0}^{\infty} E \frac{[\pm(H+1)]^n}{n!} = E q^{\pm(H+1)},
$$

$$
q^{\pm H} F = \sum_{n=0}^{\infty} \frac{(\pm H)^n}{n!} F = \sum_{n=0}^{\infty} F \frac{[\pm(H-1)]^n}{n!} = F q^{\pm(H-1)},
$$

given that $HE = E(H + 1)$ and $HF = F(H - 1)$. Therefore, we get

$$[\Delta(E), \Delta(F)] = \frac{q^{2H} - q^{-2H}}{q - q^{-1}} \otimes q^{-2H} + Fq^{H-1} \otimes Eq^{-H} - Fq^{H} \otimes Eq^{-H-1}$$

$$+ Eq^{H} \otimes Fq^{-H+1} - Eq^{H+1} \otimes Fq^{-H} + q^{2H} \otimes \frac{q^{2H} - q^{-2H}}{q - q^{-1}}$$

(8.2.6)

$$= \frac{q^{2H} - q^{-2H}}{q - q^{-1}} \otimes q^{-2H} + Fq^{H} \otimes Eq^{-H}(q^{-1} - q^{-1})$$

$$+ Eq^{H} \otimes Fq^{-H}(q - q) - Eq^{H+1} \otimes Fq^{-H} + q^{2H} \otimes \frac{q^{2H} - q^{-2H}}{q - q^{-1}},$$

where we have used the relation $[E, F] = \frac{q^{2H} - q^{-2H}}{q - q^{-1}}$, and the fact that the deformation parameter q is the same for both factors of the tensor product. Therefore, we are left with objects that all commute amongst themselves, hence

$$[\Delta(E), \Delta(F)] = \frac{q^{2H} \otimes q^{2H} - q^{-2H} \otimes q^{-2H}}{q - q^{-1}} = \frac{q^{2(H \otimes 1 + 1 \otimes H)} - q^{-2(H \otimes 1 + 1 \otimes H)}}{q - q^{-1}}$$

$$= \frac{q^{2\Delta(H)} - q^{-2\Delta(H)}}{q - q^{-1}}.$$

In the last steps we have used the property

$$q^{\pm 2(H \otimes 1 + 1 \otimes H)} = q^{\pm 2(H \otimes 1)} q^{-2(1 \otimes H)} = q^{\pm 2H} \otimes q^{\pm 2H},$$

(8.2.7)

because there is no Baker–Campbell–Haussdorff factor for commuting objects. Finally, we have proven

$$[\Delta(E), \Delta(F)] = \frac{q^{2\Delta(H)} - q^{-2\Delta(H)}}{q - q^{-1}},$$

which is the homomorphism property.

It is now clear that to generate more representations of a quantum group starting from known representations, we can tensor the latter together provided that we use the *deformed* coproduct.

Let us also remark that the case of q being a root of unity is special and the representation theory of quantum (affine) algebras is fundamentally different in that situation. This bears some relevance to the Sine-Gordon theory at special values of the coupling. We have not debated the representation theory of quantum groups at length in this review, and refer to [2–4] for a more in-depth discussion.

The antipode can easily be obtained using its defining property (8.1.16). For instance:

$$\mu(\Sigma \otimes 1)\Delta(H) = \eta\varepsilon(H) \qquad \text{hence} \qquad \Sigma(H)1 + \Sigma(1)H = 0$$
$$\text{hence} \qquad \Sigma(H) = -H.$$

and

$$\mu(\Sigma \otimes 1)\Delta(E) = \eta\varepsilon(E) \qquad \text{hence} \qquad \Sigma(E)q^{-H} + \Sigma(q^{H})E = 0$$
$$\text{hence} \qquad \Sigma(E) = -q^{-H}Eq^{H} = -q^{-1}E.$$

In the last line we have used the fact that $[H, E] = E$ implies $EH = H(E - 1)$, hence $q^{-H}Eq^{H} = q^{-H}E\sum_{n=0}^{\infty}\frac{H^{n}}{n!} = q^{-H}\sum_{n=0}^{\infty}\frac{(H-1)^{n}}{n!}E = q^{-H}q^{H-1}E$.

Exercise [10 minutes' work]: Check that the coproduct of $U_q(\mathfrak{sl}(2))$, in either of the forms that you have encountered in this book, is always coassociative.

8.3 *R*-matrix

As we have hinted at in the main text, the existence of the universal *R*-matrix can be understood in the following way. Given any $n \cdot n$ matrix representation of a Hopf algebra, the coproduct produces a new (tensor-product) representation, of dimension $n^2 \times n^2$. The latter is also the dimension of the representation obtained using the opposite coproduct. Therefore, it is possible that there exists a similarity transformation between the two. The universal *R*-matrix, if it exists, provides the formula that encodes all these similarity transformations in all of the possible representations.

Possibly using the commutation relations of $U_q(\mathfrak{g})$, one can see that, for matrix representations with nilpotent E and F, the sum in (3.6.9) truncates at $n = 1$. We can in fact bring all the Es and Fs together by paying the price of collecting more and more complicated exponents of q^H—a price that is irrelevant because any power of E and F higher than 1 then vanishes. Therefore, in particular, if $\rho(E)$ and $\rho(F)$ are nilpotent matrices, we have

$$[\rho \otimes \rho](R) = q^{2(H \otimes H)}(1 \otimes 1 + (q - q^{-1})Fq^{H} \otimes q^{-H}E). \tag{8.3.1}$$

We have surreptitiously and wittingly kept writing H, E, F instead of $\rho(H), \rho(E), \rho(F)$—hopefully with no confusion arising.

We notice that one particular representation with nilpotent E and F is the fundamental representation (3.6.11), given by

$$H = \frac{1}{2}\begin{pmatrix} 1 & 0 \\ 0 & -1 \end{pmatrix}, \qquad E = \begin{pmatrix} 0 & 1 \\ 0 & 0 \end{pmatrix}, \qquad F = \begin{pmatrix} 0 & 0 \\ 0 & 1 \end{pmatrix}. \tag{8.3.2}$$

We remind once more that it is an accident of small matrices that the fundamental representation works for both for $U_q(\mathfrak{sl}(2))$ and $\mathfrak{su}(2)$. By brute-force plugging into (8.3.1) we get

$$[\rho \otimes \rho](R) = \begin{pmatrix} \sqrt{q} & 0 & 0 & 0 \\ 0 & \frac{1}{\sqrt{q}} & 0 & 0 \\ 0 & 0 & \frac{1}{\sqrt{q}} & 0 \\ 0 & 0 & 0 & \sqrt{q} \end{pmatrix} \begin{pmatrix} 1 & 0 & 0 & 0 \\ 0 & 1 & 0 & 0 \\ 0 & (q - q^{-1}) & 1 & 0 \\ 0 & 0 & 0 & 1 \end{pmatrix} = \begin{pmatrix} \sqrt{q} & 0 & 0 & 0 \\ 0 & \frac{1}{\sqrt{q}} & 0 & 0 \\ 0 & \frac{(q - q^{-1})}{\sqrt{q}} & \frac{1}{\sqrt{q}} & 0 \\ 0 & 0 & 0 & \sqrt{q} \end{pmatrix},$$

which exactly reproduces (3.6.12).

The universal formula (3.6.9) does not only rely on inventiveness and it can actually be derived—the same holds for all simple Lie algebras and most simple Lie superalgebras. There is a theorem by Drinfeld that allows us to systematically obtain the explicit expression for the universal R-matrix (in most cases after quite a long and complicated derivation). This is based on the notion of an inner product that one puts onto the Hopf algebra, call it $\langle .|. \rangle \in \mathbb{C}$. This inner product needs to satisfy certain consistency requirements. Chief amongst these requirements is

$$\langle ab|c \rangle = \langle a \otimes b|\Delta(c) \rangle, \qquad \langle a|bc \rangle = \langle \Delta(a)|b \otimes c \rangle, \tag{8.3.3}$$

where the inner product is linear with respect to the tensor product:

$$\langle a \otimes b|c \otimes d \rangle = \langle a|c \rangle \langle b|d \rangle. \tag{8.3.4}$$

We also require

$$\langle \mathbb{1}|a \rangle = \langle a|\mathbb{1} \rangle = \varepsilon(a). \tag{8.3.5}$$

These rules allow us to define the inner product over the whole space by homomorphism. Once we have such an inner product, the universal R-matrix is constructed as

$$R = \sum_i e_i \otimes e^i, \tag{8.3.6}$$

if $\{e_i\}$ is a basis and $\{e^i\}$ is its dual basis, satisfying $\langle e^i|e_j \rangle = \delta^i_j$. The original space and its dual form the *Drinfeld quantum double*. If a Hopf algebra can be written as a double, then (8.3.6) will be the universal R-matrix. This looks like an innocuous definition, but finding the basis and then the dual basis is a formidable problem. Already in the case of a universal enveloping algebra of a Lie algebra, the space is infinite-dimensional. Defining an ordered basis in the space of all the powers of the generators is nontrivial and involves such notions as the Poincaré–Birkhoff–Witt basis. One effectively ends up solving a monumental Gram–Schmidt procedure [2–4]. In the case of (3.6.9) we can see that the factorials at the denominator have something to do with the orthonormalisation of the dual basis. We also see that by construction the universal R-matrix tends to display all the generators and all their powers, which is an awfully good way of keeping track of how 'big' the algebra is. This is particularly relevant when one is searching for extended or hidden symmetries that are not manifest from the outset—in this way, the notion of universal R-matrix has been employed for instance in AdS/CFT [12]. Notice that Drinfeld's theorem (by which we probably mean a whole collection of different theorems which Drinfeld has formulated) also guarantees the invertibility of R, plus the bootstrap conditions that we have discussed in the main text—see (3.6.13).

The notions of a particle representation ρ and an antiparticle representation $\bar{\rho}$ determine a crossing equation for the R-matrix in these specific representations. One formally has that

$$[\rho \otimes \rho](R) . [C^{-1} \otimes \mathbb{1}] . ([\bar{\rho} \otimes \rho](R))^{t_1} . [C \otimes \mathbb{1}] = \mathbb{1} \otimes \mathbb{1}, \tag{8.3.7}$$

where ^{t_1} only denotes transposition in the first factor of the tensor product. Similarly, there are analogous relations for the second factor.

8.4 RTT relations

One can show that the monodromy matrix

$$T_0 = \prod_{i=1}^{N} R_{0i} \tag{8.4.1}$$

satisfies the RTT relations (whose name is derived by the characteristic form (8.4.2)). The proof can be performed using induction, starting from the simplest case $N = 1$. We show the slightly more complicated but still very simple $N = 2$ case, to display the principle (using short-hand notation):

$$\begin{aligned} R_{00'}T_0T_{0'} &= R_{00'}R_{01}R_{02}R_{0'1}R_{0'2} = R_{00'}R_{01}R_{0'1}R_{02}R_{0'2} = R_{0'1}R_{01}R_{00'}R_{02}R_{0'2} \\ &= R_{0'1}R_{01}R_{0'2}R_{02}R_{00'} = R_{0'1}R_{0'2}R_{01}R_{02}R_{00'} = T_{0'}T_0R_{00'}, \end{aligned} \tag{8.4.2}$$

where we have used the fact that operators acting on entirely different spaces commute and then used the Yang–Baxter equation at the intermediate stages.

From an algebraic viewpoint, this means that (8.4.1) provides a representation of the RTT relations. Such a representation is, however, not the only one and the theory of representations of the RTT relations is in fact equivalent to the representation theory of the underlying quantum group.

One can also point out that if one were to take the RTT relations as abstract defining relations without having specified R first, then the Yang–Baxter equation for R would guarantee that the repeated application of the RTT relations themselves is an associative operation.

References

[1] Kassel C 1995 *Quantum Groups* (New York: Springer Science and Business Media)
[2] Chari V and Pressley A 1994 *A Guide to Quantum Groups* (Cambridge: Cambridge Univ. Press)
[3] Etingof P and Schiffmann O 2001 Lectures on the dynamical Yang–Baxter equations *Quantum Groups and Lie Theory (Durham, 1999), London Math. Soc. Lecture Note Ser.1* **290** 89–129
[4] Jimbo M 1992 Topics from representation of $U_q(g)$, *Nankai Lectures on Mathematical Physics* (Singapore: World Scientific)
[5] Loebbert F 2016 Lectures on Yangian symmetry *J. Phys.* A **49** 323002
[6] Spill F 2007 Hopf algebras in the AdS/CFT correspondence *Master's thesis* Humboldt U., Berlin
[7] Delius G W 1995 Exact S matrices with affine quantum group symmetry *Nucl. Phys.* B **451** 445–68
[8] Bernard D and Leclair A 1991 Quantum group symmetries and nonlocal currents in 2-D QFT *Commun. Math. Phys.* **142** 99–138
[9] Drinfeld V G 1989 Quasi Hopf algebras *Alg. Anal.* **1N6** 114–48
[10] Mansson T and Zoubos K 2010 Quantum symmetries and marginal deformations *JHEP* **10** 043
[11] Dlamini H and Zoubos K 2019 Marginal deformations and quasi-Hopf algebras *J. Phys.* A **52** 375402
[12] Beisert N *et al* 2012 Review of AdS/CFT integrability: an overview *Lett. Math. Phys.* **99** 3–32

IOP Publishing

Integrability using the Sine-Gordon and Thirring Duality
An introductory course
Alessandro Torrielli

Chapter 9

Supplement: Yangians

In this part of the supplement we discuss the isotropic limit of the quantum affine algebra, which goes under the name of Yangian quantum group—we will mainly follow [1].

Yangians are an extremely important class of infinite-dimensional quantum groups. In specific representations they can be associated with certain limits of quantum affine algebras, when the anisotropy characterising the quantum affine deformation tends to zero. They form the first of the three large classes of traditional quantum groups, as we will see later on when we will speak about the Belavin–Drinfeld classification.

Let us collect here the defining relations of the Yangian $\mathcal{Y}(\mathfrak{g})$ associated with a simple Lie algebra $\mathfrak{g}$. We recall that a Lie algebra $\mathfrak{g}$ is dubbed *simple* if it contains no nontrivial ideals. When working in the context of Lie algebras, an *ideal* of a Lie algebra $\mathfrak{g}$ is understood as a Lie subalgebra $\mathfrak{t} \in \mathfrak{g}$, which is such that the commutator of any element of $\mathfrak{g}$ with an element of $\mathfrak{t}$ is contained in $\mathfrak{t}$. The statement of simplicity is equivalent to the statement that the only ideals of $\mathfrak{g}$ are provided by the set $\{0\}$ and the whole algebra $\mathfrak{g}$.

There are multiple equivalent realisations of the Yangian (the same way as for quantum affine algebras). Let us recall that a *realisation* is just a different way of organising the abstract generator of an algebra (e.g., by means of linear combinations, or, in the case of the Yangian, nonlinear recombinations). A *representation* is of course something different, namely a set of maps acting on some vector space, which satisfy the abstract relations—thereby 'representing' these abstract relations via specific actions on vector spaces.

The initial realisation of the Yangian, perhaps the most natural from the physical perspective as we will later motivate, goes under the name of Drinfeld's first realisation. This realisation is rather implicit, in the sense that it only truly makes use of the level-zero and level-one generators. In fact, the Yangian can be entirely generated by these first two levels in a nonlinear fashion. Drinfeld then found another realisation (Drinfeld's second realisation) that uses all of the levels, making explicit the relations that are implicit in the first realisation. Drinfeld also gave the nonlinear map that allows us to go between the two realisations. There are excellent reviews of this formalism [2–5].

Drinfeld's first realisation of the Yangian was presented in the paper [6], and it can be seen to be rather natural from the point of view of the so-called *rational* spin-chains [7]. The very terminology of rational is in fact borrowed from the Belavin–Drinfeld classification (see later). The paper [8] very concisely announced the second realisation (in fact both the Yangians and quantum affine algebra are alike). Much work followed to perfect the proofs of the equivalence of Drinfeld's first and second realisations. It turns out that the second realisation is better suited to construct the universal *R*-matrix [9]. This is precisely because the latter makes explicit use of all the generators, as we have had the opportunity of remarking earlier. The second realisation presents a more explicit organisation of all the levels, and is therefore traditionally more convenient for the scopes of the universal *R*-matrix. There are other realisations, most notably the one provided by the RTT relations for rational *R*-matrices. The RTT realisations are of course of great importance in the study of representations by means of the algebraic Bethe ansatz [10–12]. For the reader interested in the mathematical treatment of the representation theory of Yangians and the notion of Drinfeld polynomials, we recommend [13].

We finally mention that the centre of the Yangian is associated with an object called the *quantum determinant* [11, 14], while a chosen Cartan subalgebra represents a chosen maximal Abelian subalgebra made out of the commuting Hamiltonians of the integrable system. The Hamiltonian H of any given integrable system belongs to the intersection of the centre (since H commutes with the whole Yangian worth of conserved charges) with the chosen maximal Abelian subalgebra.

9.1 Drinfeld's first realisation

In the same spirit as the universal enveloping algebras of Lie algebras, we will always define the Yangian (and likewise any other infinite-dimensional quantum group) by listing the generators and the mutual relationships that they satisfy, and considering the total space as the space of all possible products of all the generators subject to the relations (again equivalently thinking about modding the tensor algebra by the ideal generated by the relations). One calls such total space the Yangian (likewise for other infinite-dimensional quantum groups).

As a quantum group, the Yangian $\mathcal{Y}(\mathfrak{g})$ is strictly speaking a particular deformation of the universal enveloping algebra $U(\mathfrak{g}[u])$ of the *loop* algebra $\mathfrak{g}[u]$ associated with the Lie algebra $\mathfrak{g}$. As such, the Yangian is an algebra but not (the universal envelop of) a Lie algebra. The loop algebra $\mathfrak{g}[u]$ is defined as the algebra of polynomials in the complex variable u with coefficients taking values in $\mathfrak{g}$. Let us denote here the generators of the (finite-dimensional) simple Lie algebra $\mathfrak{g}$ by T^A, and let the commutation relations be $[T^A, T^B] = f_C^{AB} T^C$. Thanks to the simplicity of the Lie algebra, we have at our disposal a nondegenerate invariant bilinear form $(.,.)$. By invariant one means $([X, Y], Z) = (X, [Y, Z]) \, \forall \, X, Y, Z \in \mathfrak{g}$. Such a bilinear form is characterised by a metric κ^{AB}. The standard form that one can use is the Killing form $\kappa^{AB} = f_D^{AC} f_C^{BD}$, which is also obtained as the trace, in the adjoint representation, of the product $T^A T^B$.

The Yangian algebra can be defined in Drinfeld's first realisation by generators and relations: we have the level-zero generators T^A of $\mathfrak{g}$, and then the level-one generators $\hat{\mathfrak{J}}^A$, satisfying

$$[T^A, T^B] = f_C^{AB} T^C,$$
$$[T^A, \hat{\mathfrak{J}}^B] = f_C^{AB} \hat{\mathfrak{J}}^C. \tag{9.1.1}$$

The higher-level generators can be defined by taking successive commutation of the generators of the previous levels. This can make the algebra grow too fast. Therefore, to obtain the space that one ultimately desires, one requires a suitable set of Serre relations. Such relations put a constraint on the growth of the algebra by setting to zero certain combinations of generators. Let us specify such relations for the case when $\mathfrak{g} \neq \mathfrak{sl}(2)$:

$$[\hat{\mathfrak{J}}^A, [\hat{\mathfrak{J}}^B, T^C]] + [\hat{\mathfrak{J}}^B, [\hat{\mathfrak{J}}^C, T^A]] + [\hat{\mathfrak{J}}^C, [\hat{\mathfrak{J}}^A, T^B]]$$
$$= \frac{1}{4} f_D^{AG} f_E^{BH} f_F^{CK} f_{GHK} T^{\{D} T^E T^{F\}}. \tag{9.1.2}$$

Here the curly brackets on the indices indicate their complete symmetrisation. We also note that the indices are raised and lowered by using the metric κ^{AB} and its inverse. In the case of the Lie algebra $\mathfrak{sl}(2)$, the Serre relations (9.1.2) trivialise. One needs to require instead a different set of relations, whose description can be found, for instance, in [4]. We can see clearly here that the Yangian is not (the universal envelop of) a Lie algebra: when taking the commutator of two generators at level one we obtain an object that is naturally set at level two, but in addition we also obtain a cubic combination of generators at level-zero. We see instead the presence of a filtration: one always has to allow for tails, which start from any given level and which involve all the lower levels.

It is easy to see from (9.1.1) that there exists an (so-called *shift* of *boost*) automorphism

$$T^A \to T^A, \qquad \hat{\mathfrak{J}}^A \to \hat{\mathfrak{J}}^A + c\, T^A, \tag{9.1.3}$$

which leaves the relations (9.1.1) invariant, where c denotes any constant. It can be shown that this boost automorphism can be promoted to an automorphism of $\mathcal{Y}(\mathfrak{g})$.

The Yangian can be endowed with the structure of a Hopf algebra, by defining a coproduct that is a homomorphism of all the relations. The coproduct is specified at level zero and level one first, and then extended to the entire Yangian by homomorphism. One has

$$\Delta(T^A) = T^A \otimes 1 + 1 \otimes T^A, \tag{9.1.4}$$

$$\Delta(\hat{\mathfrak{J}}^A) = \hat{\mathfrak{J}}^A \otimes 1 + 1 \otimes \hat{\mathfrak{J}}^A + \frac{1}{2} f_{BC}^A T^B \otimes T^C, \tag{9.1.5}$$

with the typical level-zero *tail* of the coproduct at level one. As we have described in a previous section, the counit can be rather straightforwardly found and the antipode is easily obtained from the coproduct.

Exercise [15 minutes' work]: Find an appropriate counit for and derive the antipode of the level one generators of the Yangian in Drinfeld's first realisation.

9.2 Drinfeld's second realisation

The second realisation proposed by Drinfeld makes all the relations explicit by formulating them in a style that is more similar, for instance, to a Kac–Moody or a Virasoro algebra. In this spirit, the Yangian $\mathcal{Y}(\mathfrak{g})$ is generated by generators $\kappa_{i,m}$, $\xi^{\pm}_{i,m}$, $i = 1, \dots ,$ rank$\mathfrak{g}$, $m = 0, 1, 2, \dots$, subject to the relations

$$[\kappa_{i,m}, \kappa_{j,n}] = 0, \quad [\kappa_{i,0}, \xi^{\pm}_{j,m}] = \pm a_{ij}\,\xi^{+}_{j,m},$$

$$[\xi^{+}_{j,m}, \xi^{-}_{j,n}] = \delta_{i,j}\,\kappa_{j,n+m},$$

$$[\kappa_{i,m+1}, \xi^{\pm}_{j,n}] - [\kappa_{i,m}, \xi^{\pm}_{j,n+1}] = \pm\frac{1}{2}a_{ij}\{\kappa_{i,m}, \xi^{\pm}_{j,n}\},$$

$$[\xi^{\pm}_{i,m+1}, \xi^{\pm}_{j,n}] - [\xi^{\pm}_{i,m}, \xi^{\pm}_{j,n+1}] = \pm\frac{1}{2}a_{ij}\{\xi^{\pm}_{i,m}, \xi^{\pm}_{j,n}\},$$

$$i \neq j, \quad n_{ij} = 1 + |a_{ij}|, \quad Sym_{\{k\}}[\xi^{\pm}_{i,k_1}, [\xi^{\pm}_{i,k_2},\dots[\xi^{\pm}_{i,k_{n_{ij}}}, \xi^{\pm}_{j,l}]\dots]] = 0.$$

(9.2.1)

We have denoted here by a_{ij} the Cartan matrix of $\mathfrak{g}$, which is assumed to be symmetric. We can immediately see from here that the realisation (9.2.1) stems from a chosen Chevalley–Serre presentation of ordinary simple Lie algebras, to which it reduces at level zero. Notice that the presence of anticommutators $\{.,.\}$ of *bosonic* generators (we are nowhere dealing with superalgebras in this discussion) is an indication that the relations (9.2.1) define an algebra that is not (the universal envelop of) a Lie algebra.

To compare, let us reproduce the analogous realisation of the affine Kac–Moody algebra associated with $\mathfrak{g}$:

$$[T^A \otimes t^n, T^B \otimes t^m] = [T^A, T^B] \otimes t^{n+m} + (T^A, T^B)\,n\,\delta_{n,-m}\,\mathfrak{C},$$

(9.2.2)

where $\mathfrak{C}$ is a central generator.

As in the case of the Yangian, there is a derivation map which acts as

$$[d, a_n] = n\,a_{n-1},$$

(9.2.3)

for any element a_n at level n (see for instance [15]).

Both the Yangian and the Kac–Moody algebra clearly contain $\mathfrak{g}$ as a proper Lie subalgebra (setting $n = m = 0$ both in (9.2.1) and (9.2.2)). Without relying on specific representations, one can also obtain the Yangian as a certain quotient of the quantum affine Kac–Moody algebra associated with $\mathfrak{g}$—more information on this can be found in [3, 8, 16, 17].

The map between Drinfeld's two realisations is expressed by connecting the level-zero and level-one generators in the two realisations, the rest being generated accordingly. We first adopt the same Chevalley–Serre basis for $\mathfrak{g}$, with Cartan generators H_i and positive, resp. negative, root generators $E_i^{\pm}$, which is used for the second realisation. Let us indicate by $\hat{H}_i$, $\hat{E}_i^{\pm}$ the level-one generators (in the first realisation) corresponding to the chosen Chevalley–Serre decomposition. According to [8], the map is provided by

$$
\kappa_{i,0} = H_i, \quad \xi_{i,0}^{+} = E_i^{+}, \quad \xi_{i,0}^{-} = E_i^{-},
$$
$$
\kappa_{i,1} = \hat{H}_i - v_i, \quad \xi_{i,1}^{+} = \hat{E}_i^{+} - w_i, \quad \xi_{i,1}^{-} = \hat{E}_i^{-} - z_i,
$$
(9.2.4)

with

$$
v_i = \frac{1}{4} \sum_{\beta \in \Delta^+} (\alpha_i, \beta)(E_\beta^- E_\beta^+ + E_\beta^+ E_\beta^-) - \frac{1}{2} H_i^2,
$$
(9.2.5)

$$
w_i = \frac{1}{4} \sum_{\beta \in \Delta^+} \left(E_\beta^- \mathrm{ad}_{E_i^+}(E_\beta^+) + \mathrm{ad}_{E_i^+}(E_\beta^+) E_\beta^- \right) - \frac{1}{4} \{E_i^+, H_i\},
$$
(9.2.6)

$$
z_i = \frac{1}{4} \sum_{\beta \in \Delta^+} \left(\mathrm{ad}_{E_\beta^-}(E_i^-) E_\beta^+ + E_\beta^+ \mathrm{ad}_{E_\beta^-}(E_i^-) \right) - \frac{1}{4} \{E_i^-, H_i\}.
$$
(9.2.7)

We call Δ^+ the set of all the positive roots, simple and not simple. The non-simple generators are called $E_\beta^{\pm}$. They contribute to forming the Cartan–Weyl basis of $\mathfrak{g}$. The adjoint action is given by setting $\mathrm{ad}_x(y) = [x, y]$. The reader can find some more literature on the two realisations of quantum affine algebras and their mutual relationships in [18–21].

9.3 Universal R-matrix of the Yangian of $\mathfrak{su}(2)$

We can exemplify the construction of the universal R-matrix for a special Yangian, the one associated with $\mathfrak{su}(2)$ or the general case of $\mathfrak{sl}(2)$ (at the level of the algebra there is no essential distinction). We follow the construction of [9]. The formula is factorised into three main parts, suggestively corresponding to positive and negative roots, and Cartan generators. One has in fact

$$
R = R_E R_H R_F,
$$
(9.3.1)

where R_E and R_F contain the root generators and R_H the Cartan generators, in Drinfeld's second realisation. We can specify the isomorphism (9.2.4) connecting the second with the first realisation:

$$
h_0 = h, \quad e_0 = e, \quad f_0 = f,
$$
$$
h_1 = \hat{h} - v, \quad e_1 = \hat{e} - w, \quad f_1 = \hat{f} - z,
$$
(9.3.2)

with

$$
v = \frac{1}{2}(\{f, e\} - h^2), \quad w = -\frac{1}{4}\{e, h\}, \quad z = -\frac{1}{4}\{f, h\}.
$$
(9.3.3)

Drinfeld's first realisation simply states in particular

$$[h, e] = 2e, \qquad [h, f] = -2f, \qquad [e, f] = h,$$
$$[\hat{h}, e] = [h, \hat{e}] = 2\hat{e}, \qquad [\hat{h}, f] = [h, \hat{f}] = -2\hat{f}, \qquad [\hat{e}, f] = [e, \hat{f}] = \hat{h}. \tag{9.3.4}$$

We can see that there is a representation of the Yangian associated with any representation of $\mathfrak{g}$ which satisfies a certain constraint. This is the so-called *evaluation* representation, where one establishes $\hat{h} = uh$, $\hat{e} = ue$ and $\hat{f} = uf$, with u any complex (evaluation) parameter. It is important to remark that the evaluation representation sets to zero the left-hand side of the Serre relations. Therefore, only if the right-hand side is zero in the chosen representation of $\mathfrak{g}$ can such a representation be lifted to an evaluation representation of the Yangian. This is the constraint that we have just mentioned.

Utilising the map (9.3.2) in the evaluation representation, we can easily obtain the corresponding representation for the generators at level zero and one in Drinfelds' second realisation. The paper [22] shows the formula for all the levels in the representation(s) of $\mathfrak{g} = \mathfrak{su}(2)$ considered in that paper: one has in that case

$$f_n = f\left(u + \frac{h-1}{2}\right)^n,$$
$$e_n = e\left(u + \frac{h+1}{2}\right)^n, \tag{9.3.5}$$
$$h_n = ef_n - fe_n,$$

see also [9]. The generators (9.3.5) do in such particular instance(s) satisfy the relations of the second realisation (9.2.1) for the case of $\mathfrak{su}(2)$:

$$[h_m, h_n] = 0, \qquad [e_m, f_n] = h_{n+m},$$
$$[h_0, e_m] = 2\,e_m, \qquad [h_0, f_m] = -2f_m,$$
$$[h_{m+1}, e_n] - [h_m, e_{n+1}] = \{h_m, e_n\},$$
$$[h_{m+1}, f_n] - [h_m, f_{n+1}] = -\{h_m, f_n\}, \tag{9.3.6}$$
$$[e_{m+1}, e_n] - [e_m, e_{n+1}] = \{e_m, e_n\},$$
$$[f_{m+1}, f_n] - [f_m, f_{n+1}] = -\{f_m, f_n\}.$$

This includes the fundamental representation, where

$$e = \begin{pmatrix} 0 & 1 \\ 0 & 0 \end{pmatrix}, \qquad f = \begin{pmatrix} 0 & 0 \\ 1 & 0 \end{pmatrix}, \qquad h = \sigma_3, \tag{9.3.7}$$

whence one easily finds using (9.3.5)

$$e_n = \begin{pmatrix} 0 & u^n \\ 0 & 0 \end{pmatrix}, \qquad f_n = \begin{pmatrix} 0 & 0 \\ u^n & 0 \end{pmatrix}, \qquad h_n = \begin{pmatrix} u^n & 0 \\ 0 & -u^n \end{pmatrix}. \tag{9.3.8}$$

Exercise [2 hours' work]: Verify that the representation (9.3.8) satisfies all the relations (9.3.6). Then, without relying on any representation and restricting to the levels zero and one, show (therefore, purely algebraically) that the relations (9.3.6) (meaning, all those which involve at most level zero and one) are satisfied, via applying the map (9.3.2)-(9.3.3), if one knows the relations (9.3.4). You may also use that $[\hat{e}, e] = [\hat{f}, f] = 0.$

The universal R-matrix is truly defined only for the double of the Yangian, as we have motivated earlier. The Yangian is not a Drinfeld double by itself, essentially because the appropriate pairing has to match positive with *negative* levels. One therefore introduces the Yangian double just by extending the index of the level from $\mathbb{N}$ to $\mathbb{Z}$. The Yangian double can be made quasi-triangular by means of the universal R-matrix

$$R = R_E R_H R_F, \tag{9.3.9}$$

with

$$R_E = \overrightarrow{\prod_{n \geq 0}} \exp(-e_n \otimes f_{-n-1}), \tag{9.3.10}$$

$$R_F = \overleftarrow{\prod_{n \geq 0}} \exp(-f_n \otimes e_{-n-1}), \tag{9.3.11}$$

$$R_H = \prod_{n \geq 0} \exp\left\{ \mathrm{Res}_{u=v}\left[\frac{\mathrm{d}}{\mathrm{d}u}(\log H^+(u)) \otimes \log H^-(v + 2n + 1) \right] \right\}. \tag{9.3.12}$$

The meaning of the symbols is

$$\mathrm{Res}_{u=v}(A(u) \otimes B(v)) = \sum_k a_k \otimes b_{-k-1}, \tag{9.3.13}$$

if one can write $A(u) = \sum_k a_k u^{-k-1}$ and $B(u) = \sum_k b_k u^{-k-1}$. We have also introduced the generating functions

$$E^{\pm}(u) = \pm \sum_{\substack{n \geq 0 \\ n < 0}} e_n u^{-n-1}, \qquad F^{\pm}(u) = \pm \sum_{\substack{n \geq 0 \\ n < 0}} f_n u^{-n-1},$$

$$H^{\pm}(u) = 1 \pm \sum_{\substack{n \geq 0 \\ n < 0}} h_n u^{-n-1}. \tag{9.3.14}$$

Sometimes these generating functions are also called *currents*. By putting an arrow on the product one means that a specific ordering is adopted for the levels, which is described at great length in [9].

Exercise [5 days' work]: Plug into the formula for the universal R-matrix the fundamental representation (9.3.8). Call u_1 and u_2 the evaluation parameters in the two respective spaces of the tensor product. Show that the R-matrix, which you will have obtained at the end of the calculation, is a solution of the Yang–Baxter equation (you should obtain a final result proportional to $1 \otimes 1 + \dfrac{P}{u_1 - u_2}$*, where P is the permutation on quantum states, the proportionality factor being given by a rather complicated dressing factor expressible as a ratio of Gamma functions). By relying on the boost automorphism, justify why the result only depends on $u_1 - u_2$.*

Motivated by the previous exercise, we consider Yang's quantum R-matrix:

$$R = \frac{u}{u \pm 1}\left(1 \otimes 1 + \frac{P}{u}\right), \qquad u \equiv u_1 - u_2.$$

It is amusing to make a quick analysis of bound-state representations for this R-matrix. There are simple poles at $u = \mp 1$, whose residues are, respectively, proportional to $1 \otimes 1 \mp P$. These residues are clearly associated with the antisymmetric (resp., symmetric) irreducible component of the tensor-product representation (to see this, it is sufficient to see which states are annihilated by the residues).

9.4 Principal chiral model

The Yangian, in the same way as other infinite-dimensional quantum groups, has its roots in the nonlocal conservation laws that characterise integrable systems. Such conservation laws are not only those forming the tower of commuting charges but an entire tower of non-abelian ones. It is instructive to see explicitly how this works in the example of a simple classical $1 + 1$-dimensional field theory named the *principal chiral model* (PCM). The literature on nonlocal conservation laws goes far back, and it also includes attempts at quantising the associated charges by genuinely field-theoretic methods, such as point-splitting regularisation [23, 24]. The theory of quantum groups can be seen as emerging from such perturbative analysis, encapsulating in a set of operator relations the exact nonperturbative *summa* of those expansions.

The PCM is a field theory where the field is a matrix-valued function of two Minkowskian variables: $g = g(x, t)$. The matrix belongs to a simple Lie group G, which one takes to be compact and connected. The attribute of being compact is required for G to admit nontrivial finite-dimensional representations that are unitary. The Lagrangian is given by

$$\mathscr{L} = \text{tr}[\partial_\mu g^{-1}\, \partial^\mu g]. \tag{9.4.1}$$

We can see the existence of left-hand and right-hand (global) symmetries, which act as $g \to hg, gh$, respectively, where $h \in G$. The Noether currents associated with these transformations are

$$J_\mu^{L,R} = -(\partial_\mu g)g^{-1}, \, g^{-1}(\partial_\mu g), \tag{9.4.2}$$

and belong to the Lie algebra of G, call it again $\mathfrak{g}$. Let us call the generators of this Lie algebra T^A, satisfying $[T^A, T^B] = f_C^{AB}\, T^C$. This entails a decomposition of the Noether currents, and of the associated charges $\mathfrak{J}$, as follows:

$$J_\mu = J_\mu^A\, T_A, \qquad \partial^\mu J_\mu^A = 0, \qquad \mathfrak{J}^A = \int_{-\infty}^{\infty} dx\, J_0^A. \tag{9.4.3}$$

Using integration by parts, under the assumption of disregarding any boundary term, one can bring the action (9.4.1) to a form that only depends on the Noether currents (in a quadratic form). The equations of motion of the PCM coincide with the flatness condition for the currents:

$$\partial_0 J_1 - \partial_1 J_0 + [J_0, J_1] = 0. \tag{9.4.4}$$

The model admits a Lax pair, which generalises the relation (9.4.4) and introduces a spectral parameter.

Maillet brackets

The principal chiral model is an example of a system for which the Poisson structure is non-ultralocal—see the discussion below (2.1.11)—, but there is still enough structure to be able to recast it in a useful way. This is a situation where the Poisson algebra is controlled by the so-called *Maillet brackets*—a controlled generalisation of the Sklyanin brackets (2.1.6):

$$\{L_1(x, t, u), L_2(y, t, u')\} = \delta(x - y) [r_-(u, u'), L_1(x, t, u)]$$
$$+ \delta(x - y) [r_+(u, u'), L_2(x, t, u')] + \delta'(x - y) (r_-(u, u') - r_+(u, u')), \tag{9.4.5}$$

having introduced a pair of matrices (r, s) satisfying a combined relation:

$$r_+ = r + s, \qquad r_- = r - s, \tag{9.4.6}$$

$$[(r + s)_{13}(u_1, u_3), (r - s)_{12}(u_1, u_2)] + [(r + s)_{23}(u_2, u_3), (r + s)_{12}(u_1, u_2)]$$
$$+ [(r + s)_{23}(u_2, u_3), (r + s)_{13}(u_1, u_3)] = 0. \tag{9.4.7}$$

We have assumed that the matrices r, s neither depend on the fields of the theory nor carry a dependence on the space time variables. The combined relation (9.4.7) is again meant to ensure the compatibility with the Jacobi identity; however, the treatment is very delicate and involves resolving certain ambiguities [25, 26]. In particular, the brackets that are obtained for the monodromy matrix break the Jacobi identity. In [26], a specific symmetric-limit resolution of the ambiguity is proposed that restores the Jacobi identity (see [27] for a further discussion). Despite not being of the Sklyanin type, the ensuing relations still guarantee the existence of a tower of conserved quantities in involution. Quantising these brackets is challenging. In most cases where these brackets apply, such as the principal chiral model, the quantum S-matrix is known by alternative methods.

The Lax pair for the PCM reads

$$L = \frac{u\, j_0 + j_1}{u^2 - 1}, \qquad M = \frac{u\, j_1 + j_0}{u^2 - 1}. \tag{9.4.8}$$

The (r, s) matrices are known and can be found in the original literature. Even without knowing that there exists a Lax pair, and even without being able to apply the method of the monodromy matrix that we have seen at the beginning of the book, we can just use the fact that J is both conserved and flat to show the conservation $\partial^\mu \hat{J}_\mu{}^A = 0$ of these Noether currents:

$$\hat{J}_\mu{}^A = \epsilon_{\mu\nu} J^{\nu, A} + \frac{1}{2} f^A_{BC} J^B_\mu \int_{-\infty}^x dx'\, J_0^C(x'), \tag{9.4.9}$$

$$\frac{d}{dt} \hat{\mathfrak{J}}^A = \frac{d}{dt} \int_{-\infty}^{\infty} dx\, \hat{J}_0{}^A(x) = 0. \tag{9.4.10}$$

Exercise [0.5 hour's work]: Show the conservation of the nonlocal charges that have just been introduced by explicitly writing out the left-hand side of (9.4.10) and showing that all the terms cancel. You will have to use the conservation of the current J, its flatness, integration by parts and the antisymmetry of the structure constants. We also set $\epsilon_{01} = 1$.

Once the trick is learnt, one can keep nesting more and more integrals and obtain an infinite tower of such nonlocal charges. One can also say that the very existence of this tower of charges is equivalent to the fact that the model is integrable. By using the elementary Poisson brackets of the fields, one can obtain all the inequivalent Poisson brackets of the charges. This is effectively a 'classical Yangian'. Quantising these relations into the full Yangian should reproduce, bypass, or complete the field-theory work [28] into the quantum algebra of symmetries which the Yangian is.

This example allows us to make an important point about the independence of the level-one Yangian charges. If we consider the explicit expression of $\hat{\mathfrak{J}}^A$ contained in (9.4.10), then it looks like the nonlocal part might be obtained by taking the square of $\mathfrak{J}^A$ and then splitting the double integral. That is, one could perhaps rewrite the integral in (9.4.10) as $\frac{1}{2}$ of itself, plus $\frac{1}{2}$ of the same integral where x and x' have been exchanged. If this were the case, then the nonlocal part of the charge would just be dependent, although nonlinearly, on pre-existing charges. Crucially, thanks to the antisymmetry of f_{BC}^A, there is a relative minus sign between the two terms, which one obtains by splitting, and this relative minus sign prevents their recombination into the square of $\mathfrak{J}^A$. The nonlocal charges are truly independent and in the quantum theory warranted their own independent level-one generators.

It is also important to remember that there is a right-hand and a left-hand Noether current (9.4.2); therefore, in the particular case of the PCM, we have in fact two perfectly isomorphic copies of the Yangian, each containing nonlocal charges of the type that we have just been discussing.

The path-ordered exponential of the spatial part of the Lax connection, namely the monodromy matrix that we introduced at the beginning of this book, generates both the commuting and the non-commuting charges. The trace of the monodromy matrix, which we have seen gives the transfer matrix, constitutes a generating mechanism for both towers. The way to distinguish the two types of charges is typically by changing the starting point of the expansion in the spectral parameter. One could, for instance, generate the non-commuting ones by expanding at infinity. On the other hand, expanding around the origin the logarithmic derivative of the transfer matrix provides the commuting charges (the hierarchy of integrable Hamiltonians) [7, 29]. The way to determine the special point for the commuting charges is typically to set it in correspondence to the R-matrix, reducing to a projector (this point can also be different from the origin). In the case of the PCM, the reader is invited to consult [30].

There is a beautifully intuitive way to see why the coproduct, defining the Hopf-algebraic structure of the Yangian, arises naturally from the nested integral of the nonlocal charges (9.4.10). There is a treatment based on (9.4.9) and (9.4.10) to show the emergence of the coproduct from a contour-integral formulation—this is done in

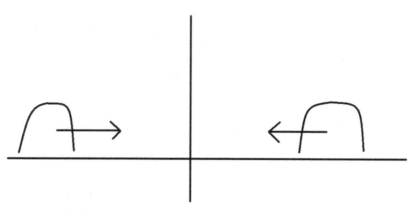

Figure 9.1. Two classical solitons heading towards each other.

[31]. A slightly more immediate way is the one provided by [23, 32]. We construct a classical solution with two solitonic profiles (which do exist in the PCM [33]) heading towards each other—see figure 9.1.

This is effectively the classical analogue of an *S*-matrix process. We imagine that the first soliton—number 1—is well localised inside the domain $(-\infty, 0)$, and soliton number 2 inside the domain $(0, \infty)$. We can classically think of the coproduct as just plugging the profile of the solution in the expression for the charges. If we do that, then we naturally have to split the integrals according to the two domains where the solitons are, respectively, localised. If we do that carefully, then we get

$$\mathfrak{J}^A_{|profile} = \int_{-\infty}^{\infty} dx \, J_{0\,|profile}^A = \int_{-\infty}^0 dx \, J_0^A + \int_0^{\infty} dx \, J_0^A$$

$$\sim \mathfrak{J}_1^A + \mathfrak{J}_2^A \longrightarrow \Delta(\mathfrak{J}^A) = \mathfrak{J}^A \otimes 1 + 1 \otimes \mathfrak{J}^A \tag{9.4.11}$$

and therefore, using (9.4.9) and (9.4.10), we see that

$$\hat{\mathfrak{J}}^A_{|profile} = \int_{-\infty}^0 dx \, J_1^A + \frac{1}{2} f^A_{BC} \int_{-\infty}^0 dx \, J_0^B(x) \int_{-\infty}^x dy \, J_0^C(y)$$

$$+ \int_0^{\infty} dx \, J_1^A + \frac{1}{2} f^A_{BC} \int_0^{\infty} dx \, J_0^B(x) \int_0^x dy \, J_0^C(y) \tag{9.4.12}$$

$$+ \frac{1}{2} f^A_{BC} \int_0^{\infty} dx \, J_0^B(x) \int_{-\infty}^0 dy \, J_0^C(y),$$

which appears to reproduce the structure of the Yangian coproduct. In fact, the first line in (9.4.12) is clearly the nonlocal charge effectively seen by the soliton to the far left, the second line is the nonlocal charge seen by the soliton to the far right, and the third line is the tail—one part left and one part right, only involving the level-zero charges as seen by the left/right soliton, respectively. Left and right are the two factors of the tensor product in this scheme, quite intuitively. We see that the structure of the Yangian coproduct, with its traditional tail, is reproduced rather immediately in this way.

9.5 More on the quantum–classical transition

To better understand how we can use the mathematics of quantum groups to formalise the quantum–classical transition in purely algebraic terms for an integrable system, let us focus on the relevant limiting procedure. For systems with a classical analogue we can take the classical limit by considering small perturbations of the quantum R-matrix around the value $1 \otimes 1$:

$$R = 1 \otimes 1 + i\,\hbar\,r + \mathcal{O}(\hbar^2), \qquad (9.5.1)$$

where $\hbar$ has been reinstated and it is typically buried within the parameters of the theory in various combinations. We also remark that the imaginary unit i conventionally appearing in (9.5.1) can in some cases be reabsorbed into a rescaling the spectral parameter u. The first nontrivial order r is identified as the classical r-matrix. The expansion (9.5.1) is interpreted as being akin to an exponential map and in some cases it has been explicitly shown to give rise to a particular very direct exponential formula [34]. In general, we have seen that the Khoroshkin–Tolstoy procedure naturally produces exponentials. According to this idea, r takes values in $\mathfrak{g} \otimes \mathfrak{g}$, with $\mathfrak{g}$ our Lie algebra, while R takes values in the tensor product of the quantum group with itself—it could be the Yangian (double), or in general a deformation of $U(\mathfrak{g}) \otimes U(\mathfrak{g})$. The classical r-matrix (just r-matrix from now on) carries a dependence on a spectral parameter, which could either scale with $\hbar$ or remain unchanged in the classical limit. To distinguish the r-matrices that do depend on a spectral parameter, and which are more interesting for integrable models, from the ones without such a dependence, the latter are dubbed 'constant' r-matrices. It is straightforward to show by Taylor expansion up to the order $\hbar^2$ that, given that R is a solution of the Yang–Baxter equation (YBE), the classical r-matrix is a solution of the classical Yang–Baxter equation (CYBE):

$$[r_{12}, r_{13}] + [r_{12}, r_{23}] + [r_{13}, r_{23}] = 0. \qquad (9.5.2)$$

The power of reducing the problem to a study and a classification of (9.5.2), which is simpler than its quantum counterpart the YBE, is that a direct link has been shown to exist between solutions of the CYBE and solutions of the quantised version. Therefore, if one is interested in classifying solutions to the YBE (which have a certain degree of regularity and standard properties—for more general cases, see for instance [35–38]), then one can to some extent focus on the classification of the solutions to the CYBE instead. The classification of the classical solutions exploits the fact that r belongs to $\mathfrak{g} \otimes \mathfrak{g}$, and is worked out in [39, 40] by means of two extremely important theorems.

- *Theorem* (First Belavin–Drinfeld theorem):

    Hypothesis: Given a simple finite-dimensional Lie algebra $\mathfrak{g}$, consider a given solution $r(u_1, u_2)$ of the CYBE belonging to $\mathfrak{g} \otimes \mathfrak{g}$. Assume such a solution to be already in difference form, namely $r(u_1, u_2) = r(u_1 - u_2)$. Suppose that any one of these three equivalent conditions hold: (i) $r(\delta u)$ has at least a pole in $\delta u = u_1 - u_2$ and moreover there exists no Lie subalgebra $\mathfrak{g}' \subset \mathfrak{g}$ such that $r(\delta u) \in \mathfrak{g}' \otimes \mathfrak{g}'$ for all δu; (ii) $r(\delta u)$ has a pole at $\delta u = 0$, the residue being proportional to the element $\sum_a I_a \otimes I_a$, where by I_a we have denoted a basis of $\mathfrak{g}$ which is orthonormal with respect to a specified nondegenerate invariant bilinear form—this could be the Killing form, and the residue so identified is effectively the quadratic Casimir $C_\otimes$ of the tensor algebra $\mathfrak{g} \otimes \mathfrak{g}$; (iii) if one considers the matrix $r_{ab}(\delta u)$ constructed using the entries (coefficients) r_{ab} of the r-matrix $r(\delta u) = \sum_{a,b} r_{ab}(\delta u) I_a \times I_b$, then this matrix is not identically degenerate (has non-everywhere-zero determinant).

Thesis: Unitarity (skew-symmetry) is satisfied, namely we have $r_{12}(\delta u) = -r_{21}(-\delta u)$. The r-matrix $r(\delta u)$ can be continued to be a meromorphic function in δu. The singularities of $r(\delta u)$ are all simple poles; these poles form a lattice Γ in the complex plane. Modulo transformations by automorphisms (see [39] for the precise statement), the matrix $r(\delta u)$ can only come in three types: if Γ is a two-dimensional lattice, then we have the elliptic case; if Γ is a one-dimensional lattice, then we have the trigonometric case; and if $\Gamma = \{0\}$ (zero-dimensional lattice), then we have the rational case (related to the types of poles presented by elliptic, trigonometric and rational functions, respectively).

The proof is rather involved and can be found in the original papers. In particular, it is quite not-trivial to demonstrate the equivalence of the three statements contained in the hypothesis. Notice that the second of these statement implies that, in a sense, the r-matrix uses all the available generators, the same as what we saw the universal R-matrix also doing. If this is not the case, i.e., if the subalgebra $\mathfrak{g}'$ mentioned in the hypothesis of the theorem does exist, then we can just restrict to this subalgebra for the hypothesis to be verified. Therefore, we can see that both the classical r-matrix and the quantum R-matrix are a very good way to figure out the exact extent of the symmetry algebra, by forcing one all the way down until the really essential set of generators is identified, but no less (see also later on, when we will discuss the two span subalgebras).

The classification is then extended from the r-matrix to the quantum R-matrix, going through the process of promoting a classical to a quantum Lie bialgebra. We refer to [3] and references quoted therein for a description of this procedure. We see therefore that the notion of quantisation acquires a very direct algebraic sense, in which one is to promote a classical r-matrix and the associated algebraic structure to the R-matrix of a quantum group, obtaining in this way a solution of the Yang–Baxter equation. As we have already had the opportunity of pointing out, and as will be even more explicit in our next example, this idea is encapsulated in the following correspondence:

$$\{A, B\} = \lim_{\hbar \to 0} \frac{[A, B]}{-i\,\hbar}. \tag{9.5.3}$$

We can therefore pinpoint a precise analytic understanding of (9.5.3) as a function of $\hbar$ [41], and reduce the quantum–classical transition to the problem of taking the limit of a matrix-valued function.

The types of quantum groups that are generated by the quantisation procedure, which we have just alluded to, are in a very general way classified as follows: elliptic quantum groups arise from the case $dim(\Gamma) = 2$, trigonometric quantum groups arise from the case $dim(\Gamma) = 1$, while Yangians arise from the case $\Gamma = \{0\}$. In the case of the Yangian, there is a very neat and intuitive way to see the emergence of the quantum group relations from the classical structure, exemplified by the prototypical case of the so-called *Yang's r-matrix* [42]:

$$r = \frac{C_\otimes}{u_2 - u_1},$$
(9.5.4)

for any Lie algebra $\mathfrak{g}$. It can be shown that this is a solution of the CYBE: one exploits the fact that $C_\otimes$ is a tensor-product Casimir, i.e., $[C_\otimes, \mathfrak{J}^A \otimes \mathbb{1} + \mathbb{1} \otimes \mathfrak{J}^A] = 0 \; \forall A$. Now what one does is to rewrite the r-matrix in terms of a geometric expansion:

$$r = \frac{C_\otimes}{u_2 - u_1} = \frac{T^A \otimes T_A}{u_2 - u_1} = \sum_{n \geqslant 0} T^A u_1^{\,n} \otimes T_A u_2^{-n-1} \equiv \sum_{n \geqslant 0} T_n^A \otimes T_{A,-n-1},$$
(9.5.5)

valid in the domain $|u_1/u_2| < 1$. The expansion in the complementary domain corresponds to swapping the two copies within the Yangian double, see the comments below (9.5.6). Here we have employed the quadratic form κ_{AB} to write $C_\otimes$ using the generators $T^A \in \mathfrak{g}$. It is clear that the geometric expansion is necessary because the spectral parameter u_1 (resp., u_2) has to belong to the first (resp., the second) space in the tensor product. In this way we have achieved for the r-matrix a precise rendering as an element of the tensor-product algebra, which is ready for interpretation. It suggests the definition $T_n^A \equiv u^n \, T^A$ in (9.5.5), in such a way that we see the emergence of a loop algebra via the following relations,

$$[T_m^A, T_n^B] = f_C^{AB} \, T_{m+n}^C,$$
(9.5.6)

which naturally follow from the definition that we have just posed. It is actually possible to explicitly prove, only using (9.5.6), that the expression $r = \sum_{n \geqslant 0} T_n^A \otimes T_{A,-n-1}$, see (9.5.5), satisfies the CYBE universally, i.e., without having to specify any representation.

Exercise [3 hours' work]: Prove the statement that has just been made.

The Yangian is then a natural quantisation (deformation) of the loop algebra that we have in this way uncovered as sitting within Yang's r-matrix. The interpretation of Yangians as deformations of loop algebras is something that we have mentioned earlier in the book, but which is here directly put in correspondence with the quantisation of Yang's r-matrix (see also later on for an explicit field-theoretic realisation of this fact).

More explicitly, we can perform a simple exercise using the CYBE, to clarify even further the emergence of the Yangian. We in fact decompose again as usual $r(\delta u) = r_{ab}(\delta u) \, I^a \otimes I^b$, and expand the left-hand side of the CYBE near $u_1 = u_2$.

Since we know that under the assumptions of the Belavin–Drinfeld theorem, r_{12} shall have a pole there, we obtain at the leading order

$$\frac{c_{ab}(u_1)}{u_1 - u_2}\, r_{cd}(u_1 - u_3)\, ([I^a, I^c] \otimes I^b \otimes I^d + I^a \otimes [I^b, I^c] \otimes I^d) = 0.$$

We have denoted by $c_{ab}(u_1)I^a \otimes I^b$ the residue of the classical r-matrix at the pole. We can see that a consequence of this necessarily is

$$[I^a, I^c] = f_m^{ac} I^m$$

for some choice of constants f_m^{ac}, which is a sufficient condition to conclude that the spans of the respective factors of r form Lie subalgebras. One can then prove [39] that the Jacobi identity is satisfied. In this way we have that the spans of the two operatorial parts, in which the classical r-matrix is split (exactly like in the case described above of Yang's r-matrix), form two Lie subalgebras of $\mathfrak{g}$. We call them 'span' subalgebras because they are obtained via the span of the two factors of the r-matrix, respectively.

It turns out that the specification of the two 'span' subalgebras provides a fingerprint for the algebraic structure underlying a given r-matrix. We have seen this quite clearly for the situation of Yang-'s r-matrix because the two spans can be put into correspondence via (9.5.5) with two identical copies of the loop algebra, basically $\mathfrak{g}[u_1] \otimes u_2^{-1}\mathfrak{g}[u_2^{-1}]$. Here we again denote by $\mathfrak{g}[u]$ the loop algebra, which we remind is the Lie algebra of $\mathfrak{g}$-valued polynomials in u (the latter being a complex variable). It is known that the loop algebra is an undeformed (classical) limit of the corresponding Yangian $\mathcal{Y}(\mathfrak{g})$, and, vice versa, that the Yangian is a quantisation, or deformation, of the loop algebra, as we have mentioned in section 9.1. One particularity of the Yangian is that once the deformation is turned on, it does not explicitly depend on the exact scale of $\hbar$, which can be reabsorbed in the generators. The presence of the deformation is, however, clear if one looks for instance at the right-hand side of the Serre relations, with the typical nonlinear deformation term.

As we have seen early on in this book, the classical r-matrix is an essential object in an integrable system with classical analogue, given that it enters the Poisson relations of the entries of the L-matrix in the classical inverse scattering method via the Sklyanin brackets [3, 16, 17, 41, 43]. It is therefore very exciting to be able to directly connect this cardinal piece of the classical picture to the quantum group controlling the quantum problem. We will see exactly how this connects in the next section, in the particular case of the Lieb–Liniger model, to what is normally meant in physics by quantisation of a classical system (in that case it will be via the normal-ordering prescription).

Let us also mention another interesting mathematical fact. The assumption of difference form is not too restrictive, thanks to the following theorem by Belavin and Drinfeld [44]:

Theorem (Second Belavin–Drinfeld theorem):

Hypothesis: Assume the very same hypothesis as in the first Belavin–Drinfeld theorem, except for the fact that we do not assume $r = r(u_1, u_2)$ to be of difference form. The CYBE is written very naturally in this case as

$$[r_{12}(u_1, u_2), r_{13}(u_1, u_3)] + [r_{12}(u_1, u_2), r_{23}(u_2, u_3)] + [r_{13}(u_1, u_3), r_{23}(u_2, u_3)] = 0. \quad (9.5.7)$$

The three statements that were indicated by (i), (ii) and (iii) in the first Belavin–Drinfeld theorem are no longer equivalent, so we now need to choose one; the standard assumption is to maintain (ii) as a part of the hypothesis. The simplicity of the Lie algebra has the effect that the dual Coxeter number is nonzero. We recall our definition of the dual Coxeter number, which we call c_2: we choose to define $\sum_{ab} f_{abc} f_{abd} = c_2 \delta_{cd}$. This computes the trace of the Casimir when taken in the adjoint representation, i.e., $\sum_a [T_a, [T_a, x]] = -c_2 x, \ \forall \ x \in \mathfrak{g}$. We can also express the Killing form of $\mathfrak{g}$ as $\kappa_{cd} = \sum_{ab} f_{abc} f_{abd} = c_2 \delta_{cd}$, and it will be nondegenerate for a simple Lie algebra.

Thesis: One can find a transformation (which includes a change a variables) that sends $r(u_1, u_2)$ to being of difference form.

In this case the proof is more manageable and we can report it, see also [27]. The proof proceeds as follows. Let us assume for the r-matrix the following behaviour near the pole at $u_1 = u_2$:

$$r \sim \frac{\sum_a T_a \otimes T_a}{u_1 - u_2} + g(u_1, u_2). \tag{9.5.8}$$

This is not reductive and it brings to no loss of generality. In fact, if our assumption were not verified, then we would have

$$r \sim \frac{\xi(u_1) \sum_a T_a \otimes T_a}{u_1 - u_2} + g(u_1, u_2). \tag{9.5.9}$$

With the simple change of variables

$$u = u(v), \qquad u'(v) = \xi(u(v)), \tag{9.5.10}$$

we can achieve the desired form of the residue. To see this, we can expand around $v_1 = v_2$ and obtain

$$\frac{\xi(u_1)}{u_1 - u_2} \sim \frac{\xi(u(v_1))}{u(v_2) + u'(v_2)(v_1 - v_2) - u(v_2)} \sim \frac{1}{v_1 - v_2}, \tag{9.5.11}$$

having considered (9.5.10). We can assume that g is analytic in a vicinity of $v_1 = v_2$. The key is then to perform an expansion of (9.5.7) around $u_2 = u_3$, to obtain

$$[r_{12}(u_1, u_2), r_{13}(u_1, u_2)] + [r_{12}(u_1, u_2) + r_{13}(u_1, u_2), g_{23}(u_2, u_2)]$$

$$+ \left[\sum_{ab} r_{ab}(u_1, u_2) T_a \otimes \Delta(T_b), 1 \otimes \frac{\sum_c T_c \otimes T_c}{u_2 - u_3} \right] \tag{9.5.12}$$

$$+ \left[\partial_{u_2} r_{12}(u_1, u_2), 1 \otimes \sum_a T_a \otimes T_a \right] = 0.$$

The coproduct $\Delta(T_a) = T_a \otimes 1 + 1 \otimes T_a$ is the standard coproduct on $U(\mathfrak{g})$. Therefore, we see that the third contribution in the above formula cancels out due to the fact that the tensor Casimir $\sum_a T_a \otimes T_a$ commutes with the standard coproduct: $[\Delta(T_b), \sum_a T_a \otimes T_a] = 0$ (as in the ordinary quantum mechanics of the spin).

Following Belavin and Drinfeld, we consider the map which is the commutator itself: $x \otimes y \rightarrow [x, y]$. We can apply this map to the second and third space of the equation (9.5.12) that we have just found. We can also rely on the nonvanishing of the dual Coxeter number, which we can set to -1 by adopting suitable conventions. By use of the Jacobi identity, we can recast the equation as

$$\sum_{abcd} r_{ab}(u_1, u_2) r_{cd}(u_1, u_2)[T_a, T_c] \otimes [T_b, T_d] + [r(u_1, u_2), 1 \otimes h(u_2)] + \partial_{u_2} r(u_1, u_2) = 0,$$

having posed $h(u) \equiv g_{ab}(u, u)[T_a, T_b]$.

By doing the same on u_1 and u_2 and applying the same map to the first and second space, we can likewise write

$$\sum_{abcd} r_{ab}(u_1, u_3) r_{cd}(u_1, u_3)[T_a, T_c] \otimes [T_b, T_d] + [h(u_1) \otimes 1, r(u_1, u_3)] - \partial_{u_1} r(u_1, u_3) = 0.$$

In the end we can subtract the two equations that we have obtained, re-evaluating both at u_1 and u_2, resulting in

$$\partial_{u_1} r(u_1, u_2) + \partial_{u_2} r(u_1, u_2) = [h(u_1) \otimes 1 + 1 \otimes h(u_2), r(u_1, u_2)]. \tag{9.5.13}$$

The last step is to introduce a map $\psi(u)$ acting, in an invertible way, on $\mathfrak{g}$, defined by

$$\frac{d}{du} \psi(u)[x] = [h(u), \psi(u)[x]] \qquad \forall \ x \in \mathfrak{g}. \tag{9.5.14}$$

We can also define

$$\hat{r}(u_1, u_2) = [\psi^{-1}(u_1) \otimes \psi^{-1}(u_2)] r(u_1, u_2). \tag{9.5.15}$$

Given that

$$\frac{d}{du} \psi^{-1}(u) = -\psi^{-1}(u) \left(\frac{d}{du} \psi(u) \right) \psi^{-1}(u), \tag{9.5.16}$$

(9.5.13) can be recast as

$$\partial_{u_1} \hat{r}(u_1, u_2) + \partial_{u_2} \hat{r}(u_1, u_2) = 0, \tag{9.5.17}$$

where we have used that in particular the map $\frac{d}{du_i} \psi(u_i)$, $i = 1, 2$, acts on everything that appears to its right (in its space i of the tensor product). But this clearly shows the difference form of $\hat{r}$, and we have therefore proven the assertion that we set out to prove.

References

[1] Torrielli A 2011 Yangians, S-matrices and AdS/CFT *J. Phys.* A **44** 263001

[2] Loebbert F 2016 Lectures on Yangian symmetry *J. Phys.* A **49** 323002

[3] Etingof P and Schiffmann O 2001 Lectures on the dynamical Yang–Baxter equations *Quantum Groups and Lie Theory (Durham, 1999), London Math. Soc. Lecture Note Ser* **290** 89–129

[4] MacKay N J 2005 Introduction to Yangian symmetry in integrable field theory *Int. J. Mod. Phys.* A **20** 7189–218

[5] Molev A, Nazarov M and Olshansky G 1996 Yangians and classical Lie algebras *Russ. Math. Surv.* **51** 205

[6] Drinfeld V G 1986 Quantum groups *Zap. Nauchn. Semin.* **155** 18–49

[7] Bernard D 1993 An introduction to Yangian symmetries *Int. J. Mod. Phys.* B **7** 3517–30

[8] Drinfeld V G and New A 1988 A new realization of Yangians and quantized affine algebras *Sov. Math. Dokl.* **36** 212–6

[9] Khoroshkin S M and Tolstoy V N 1996 Yangian double *Lett. Math. Phys.* **36** 373–402

[10] Faddeev L D, Reshetikhin N Y and Takhtajan L A 1989 Quantization of Lie groups and Lie algebras *Alg. Anal.* **1** 178–206

[11] Kirillov A N and Reshetikhin N Y 1986 Bethe ansatz and combinatorics of Young tableaux *Zap. Nauchn. Sem. POMI* **155** 65–115

[12] Kirillov A N and Reshetikhin N Y 1990 Representations of Yangians and multiplicities of occurrence of the irreducible components of the tensor product of representations of simple Lie algebras *J. Sov. Math.* **52** 3156–64

[13] Pressley A and Chari V 1991 Fundamental representations of Yangians and singularities of R-matrices *J. Reine Angew. Math.* **417** 87–128

[14] Kulish P P and Sklyanin E K 1982 Quantum spectral transform method: recent developments *Lect. Notes Phys.* **151** 61–119

[15] Goddard P and Olive D I 1986 Kac-Moody and Virasoro algebras in relation to quantum physics *Int. J. Mod. Phys.* A **1** 303

[16] Chari V and Pressley A 1994 *A Guide to Quantum Groups* Cambridge: Cambridge Univ. Press

[17] Jimbo M 1992 Topics from representation of Uq(g) *Nankai Lecture Series in Mathematical Physics* (Singapore: World Scientific) https://doi.org/10.1142/1469

[18] Khoroshkin S M and Tolstoy V N 1993 On Drinfeld's realization of quantum affine algebras *J. Geom. Phys.* **11** 445–52

[19] Ding J and Frenkel I B 1993 Isomorphism of two realizations of quantum affine algebra *Commun. Math. Phys.* **156** 277–300

[20] Jing N 1998 On Drinfeld realization of quantum affine algebras *Ohio State Univ. Math. Res. Inst. Publ. de Gruyter, Berlin* **7** 195–206

[21] Hayaishi N and Miki K 1998 L operators and Drinfeld's generators *J. Math. Phys.* **39** 1623–36

[22] Arutyunov G, de Leeuw M and Torrielli A 2009 Universal blocks of the AdS/CFT scattering matrix *JHEP* **05** 086

[23] Luscher M and Pohlmeyer K 1978 Scattering of massless lumps and nonlocal charges in the two-dimensional classical nonlinear sigma model *Nucl. Phys.* B **137** 46–54

[24] Brezin E, Itzykson C, Zinn-Justin J and Zuber J B 1979 Remarks about the existence of nonlocal charges in two-dimensional models *Phys. Lett.* B **82** 442–4

[25] Maillet J M 1986 Hamiltonian structures for integrable classical theories from graded Kac–Moody algebras *Phys. Lett.* B **167** 401–5

[26] Maillet J M 1986 New integrable canonical structures in two-dimensional models *Nucl. Phys.* B **269** 54–76

[27] Torrielli A 2016 Lectures on classical integrability *J. Phys.* A **49** 323001

[28] Luscher M 1978 Quantum nonlocal charges and absence of particle production in the two-dimensional nonlinear sigma model *Nucl. Phys.* B **135** 1–19

[29] Tarasov V O, Takhtajan L A and Faddeev L D 1983 Local Hamiltonians for integrable quantum models on a lattice *Theor. Math. Phys.* **57** 1059–73

[30] Evans J M, Hassan M, MacKay N J and Mountain A J 1999 Local conserved charges in principal chiral models *Nucl. Phys.* B **561** 385–412

[31] Bernard D and Leclair A 1991 Quantum group symmetries and nonlocal currents in 2-D QFT *Commun. Math. Phys.* **142** 99–138

[32] MacKay N J 1993 On the classical origins of Yangian symmetry in integrable field theory *Phys. Lett.* B **281** 90–7 [Erratum: Phys. Lett. B 308, 444–4 (1993)]

[33] Zakharov V E and Mikhailov A V 1978 Example of nontrivial interaction of solitons in two-dimensional classical field theory *JETP Lett.* **27** 42–6

[34] Freidel L and Maillet J M 1992 The universal R matrix and its associated quantum algebra as functionals of the classical R matrix: the sl(2) case *Phys. Lett.* B **296** 353–60

[35] de Leeuw M, Paletta C, Pribytok A, Retore A L and Ryan P 2020 Classifying nearest-neighbor interactions and deformations of AdS *Phys. Rev. Lett.* **125** 031604

[36] de Leeuw M, Paletta C, Pribytok A, Retore A L and Ryan P 2021 Yang–Baxter and the Boost: splitting the difference *SciPost Phys.* **11** 069

[37] Pribytok A 2021 Automorphic symmetries and AdSn integrable deformations *In 14th International Workshop on Lie Theory and Its Applications in Physics* **12**

[38] Pribytok A 2022 Automorphic symmetries, string integrable structures and deformations arXiv:2210.16348 [hep-th]

[39] Belavin A A and Drinfeld V G 1982 Solutions of the classical Yang–Baxter equation for simple Lie algebras *Funktsional'nyi Analiz i ego Prilozheniya* **16** 1–29

[40] Belavin A A and Drinfeld V G 1984 *Triangle equations and Simple Lie Algebras* (New York: Harwood Academic)

[41] Sklyanin E K 1980 Quantum version of the method of inverse scattering problem *Zap. Nauchn. Semin.* **95** 55–128

[42] Yang C-N 1967 Some exact results for the many body problems in one dimension with repulsive delta function interaction *Phys. Rev. Lett.* **19** 1312–4

[43] Faddeev L D and Takhtajan L A 1987 *Hamiltonian Methods in the Theory of Solitons* (Berlin: Springer)

[44] Belavin A A and Drinfeld V G 1983 Classical Young–Baxter equation for simple lie algebras *Funktsional'nyi Analiz i ego Prilozheniya* **17** 69–70

IOP Publishing

Integrability using the Sine-Gordon and Thirring Duality
An introductory course
Alessandro Torrielli

Chapter 10

Supplement: the Lieb–Liniger model

We shall now describe in some detail the quantisation of the Lieb–Liniger model, also known as the nonlinear Schroedinger equation. The purpose of this exercise is to exemplify many of the concepts that we have encountered in studying the Sine-Gordon/Thirring model, working in a much simpler and clearer (though less rich) setup. The more straightforward links with perturbation theory and the simplicity of the formulas will be of great benefit if one then wishes to return to the Sine-Gordon model with a fresh eye.

10.1 The classical theory

We shall start with the classical Hamiltonian

$$H_{LL} = \int_{-\infty}^{\infty} dx \left[\frac{\partial \psi^*}{\partial x} \frac{\partial \psi}{\partial x} + \kappa \, \psi^{*2} \psi^2 \right] \tag{10.1.1}$$

for a complex scalar field $\psi(x)$ in $1 + 1$ dimensions, with coordinates (x, t), enjoying Galileian symmetry, κ being a real coupling constant. We denote by $\psi^*(x)$ the complex conjugate of $\psi(x)$ (and will also systematically suppress the time dependence in the argument of the fields). The equal-time Poisson brackets read

$$\{\psi(x), \psi^*(y)\} = i\delta(x - y). \tag{10.1.2}$$

This system is actually a constrained system because it has a Lagrangian density that is 'linear in the velocities':

$$L_{LL} = i \, \psi^* \frac{\partial \psi}{\partial t} - \frac{\partial \psi^*}{\partial x} \frac{\partial \psi}{\partial x} - \kappa \, \psi^{*2} \psi^2. \tag{10.1.3}$$

This means that imposing $P_\psi \equiv \frac{\delta L_{LL}}{\delta \partial_t \psi}$ as a definition of the conjugate momentum results in a constraint: $P_\psi = i\psi^*$. Basically, one cannot eliminate $\partial_t \psi$ in favour or Π_ψ

using the defining equation of the conjugate momentum. The treatment of a constrained system is most rigorously performed using Dirac's theory of constrained Hamiltonian systems: in this case this is very clearly done, for instance, in section 3.1 of [1]. The Poisson structure (10.1.2) can be seen as emerging from the Dirac brackets of the constrained procedure.

The equations of motion

$$\frac{d\psi}{dt} = \{H_{LL}, \psi\} \tag{10.1.4}$$

imply

$$i\frac{\partial\psi}{\partial t} = -\frac{\partial^2\psi}{\partial x^2} + 2\kappa |\psi|^2\psi, \tag{10.1.5}$$

whence the name *nonlinear Schroedinger equation*. In fact, (10.1.5) is a Schroedinger equation in one spatial dimension, with a potential that depends nonlinearly on the wave function itself.

The Lieb–Liniger model is classically (and quantum-mechanically) integrable. A Lax pair for the system is given by equation (3.24) in [2]. The monodromy matrix is also reported in [2]—equation (3.25) and two un-numbered equations following that, where one can see the entries expressed in a perturbative expansion in the coupling constant κ. We report them here for convenience (correcting a typo in [2]):

$$a(u) = e^{-i\frac{u}{2}(s_+-s_-)}\left[1 + \sum_{n=1}^{\infty}\kappa^n \int_{s_+>\xi_n>\eta_n>\xi_{n-1}\cdots>\eta_1>s_-} d\xi_1\cdots d\xi_n\, d\eta_1\cdots d\eta_n\right.$$
$$\left. e^{iu(\xi_1 + \cdots+\xi_n-\eta_1-\cdots-\eta_n)}\psi^*(\xi_1)\cdots\psi^*(\xi_n)\,\psi(\eta_1)\cdots\psi(\eta_n)\right],$$

$$b(u) = -i\,e^{i\frac{u}{2}(s_+ + s_-)}\left[\sum_{n=0}^{\infty}\kappa^n \int_{s_+>\eta_{n+1}>\xi_n>\eta_n>\xi_{n-1}\cdots>\eta_1>s_-} d\xi_1\cdots d\xi_n\, d\eta_1\cdots d\eta_{n+1}\right.$$
$$\left. e^{iu(\xi_1 + \cdots+\xi_n-\eta_1-\cdots-\eta_{n+1})}\psi^*(\xi_1)\cdots\psi^*(\xi_n)\,\psi(\eta_1)\cdots\psi(\eta_{n+1})\right].$$

The theory is put on a finite interval $x \in [s_-, s_+]$ with periodic boundary conditions. A suitably adapted expression holds in the infinite-interval case [3, 4].

The first few conserved charges (on the infinite interval $s_- \to -\infty$, $s_+ \to \infty$) are also displayed in [2]—equation (3.27). It is important to note that the trace of the monodromy matrix (namely the transfer matrix) is real for real values of the spectral parameter u. It is also important to observe that the perturbative expansions of the entries of the monodromy matrix see a balance between ψs and ψ^*s in the diagonal entries (the functionals a and a^*) and an excess (defect, respectively) of one ψ^* versus a ψ in the entry b^* (b, respectively). This will become relevant when quantising these objects because they will preserve and create (annihilate, respectively) particles in the quantum theory.

10.2 Quantisation

The quantisation of the Lieb–Liniger model follows, for instance, from [4]. The reader can also consult [5] (e.g., their section 3) for general considerations on the quantisation of systems that are of the first order in the time derivative. One starts by replacing the classical variables with quantum operators

$$\psi(x) \rightarrow \psi(x), \qquad \psi^*(x) \rightarrow \psi^\dagger(x), \qquad (10.2.1)$$

with commutation relations

$$[\psi(x), \psi^\dagger(y)] = \delta(x - y). \qquad (10.2.2)$$

One then postulates the existence of a perturbative vacuum $|0\rangle$ defined by the condition

$$\psi(x)|0\rangle = 0 \qquad \forall\, x \in \mathbb{R}. \qquad (10.2.3)$$

In this theory, it will turn out that the perturbative vacuum is also the exact nonperturbative vacuum of the theory. On the other hand, repeatedly acting with $\psi^\dagger(x)$ creates perturbative excitations (one $\psi^\dagger(x)$ creates a perturbative particle in position x), but it does not create exact eigenstates of the Hamiltonian. The correct nonperturbative creation operators are much more complicated than $\psi^\dagger$. Let us see how it works.

10.2.1 Direct diagonalisation

We start in infinite volume. The first thing to consider is that the exact eigenstates of the quantum Hamiltonian can, in this simpler problem, be found by brute force. This is thanks to the conservation of the particle number

$$\mathcal{N} = \int_{-\infty}^{\infty} dx\; \psi^\dagger(x)\psi(x). \qquad (10.2.4)$$

This operator commutes with the quantum Hamiltonian, which is quantised by *normal ordering*, namely setting all the $\psi^\dagger$s to the left of the ψs.

This is a particularly fortunate case because of the following reason. Although the perturbative particles created by $\psi^\dagger$ are not the exact nonperturbative particles, the number operator (10.2.4) is nevertheless conserved exactly. This means that we can still label the exact eigenstates of H_{LL} by the number of perturbative excitations that they contain. This allows us to effectively break down the quantum field theory into a series of quantum mechanical problems, each taken at a fixed natural value M of the spectrum of $\mathcal{N}$. In each of these sectors one reduces the problem to the following Schroedinger's equation:

$$\left[-\sum_{i=1}^{M} \frac{\partial^2}{\partial x_i^2} + 2\kappa \sum_{i<j=1}^{M} \delta(x_i - x_j) - E \right] f_M(x_1, \ldots, x_M) = 0, \qquad (10.2.5)$$

where the wave function f_M is defined by

$$H|f_M\rangle = E|f_M\rangle,$$

$$|f_M\rangle = \frac{1}{\sqrt{M!}} \int_{-\infty}^{\infty} dx_1 \cdots \int_{-\infty}^{\infty} dx_M \, f_M(x_1, \ldots, x_M) \psi^\dagger(x_1) \cdots \psi^\dagger(x_M)|0\rangle \qquad (10.2.6)$$

(with $|0\rangle$ being the perturbative, and in this particular case also nonperturbative, vacuum state). Notice that the wave function $f_M(x_1, \ldots, x_M)$ can be chosen to be completely symmetric in its arguments, due to the commutativity of the bosonic operators $\psi^\dagger$. Formula (10.2.5) shows how this problem is in fact equivalent to a *one-dimensional Bose gas* with delta-function potential, namely a gas of bosons interacting only when their locations pairwise coincide.

The fact that $|f_M\rangle$ is an eigenstate of H_{LL} if (10.2.5) is satisfied can be proven by brute-force plugin and repeated use of the fundamental commutation relations (10.2.2). All that is left to do is to solve the Schroedinger problem, and this can also be done by brute force. The solution is given by the following expression in the hyper-wedge $x_1 < \cdots < x_M$:

$$f_M(x_1, \ldots, x_M) = \frac{i^M}{(2\pi)^{\frac{M}{2}} \sqrt{M!}} \sum_{\sigma \in S_M} c_{a_{\sigma(1)} \ldots a_{\sigma(M)}} \exp i\left[p_{\sigma(1)} x_1 + \cdots + p_{\sigma(M)} x_M \right], \qquad (10.2.7)$$

and then extended by symmetry everywhere else. The coefficients $c_{a_1 \ldots a_M}$ satisfy in particular

$$\frac{c_{a_1 \ldots a_{j-1} a_j \ldots a_M}}{c_{a_1 \ldots a_j a_{j-1} \ldots a_M}} = \frac{p_{j-1} - p_j + i\kappa}{p_{j-1} - p_j - i\kappa}, \qquad (10.2.8)$$

where the right-hand side of (10.2.8) is a pure phase for real momenta and coupling. When all the momenta (and the coupling κ) are real, the choice (2.1.13) in [4], namely

$$c_{a_1 \ldots a_M} = \prod_{r<s} \sqrt{\frac{p_{a_r} - p_{a_s} + i\kappa}{p_{a_r} - p_{a_s} - i\kappa}}, \qquad (10.2.9)$$

guarantees that the wave function is delta-function normalised. The spectrum is therefore labelled by sets of momenta $\{p_1, \ldots, p_M\}$, and the wave function has the typical scattering form of a superposition of plane waves with all the possible permutation of the momenta and coefficients which are related to one another by factors of the fundamental S-matrix

$$S_{12} = S(p_1, p_2) = S_{LL}(p_1 - p_2) = \frac{p_1 - p_2 + i\kappa}{p_1 - p_2 - i\kappa}. \qquad (10.2.10)$$

The Galilei invariance of the problem is reflected in the dependence of the S-matrix on the difference of the momenta (according to the Galileian formula of linear addition of the velocities). The energy of the state $|f_M\rangle$ is simply

$$E = \sum_{i=1}^{M} p_i^2 \qquad (10.2.11)$$

independent on the coupling κ. In the case of two particles $M = 2$, one can verify by simple manipulations that the above wave function is proportional to (3.3.17).

Let us show the phenomenon of fermionisation explicitly for the case of three particles. We can write

$$f_3(x_1, x_2, x_3) = \frac{i^3}{(2\pi)^{\frac{3}{2}}\sqrt{6}} \left[e^{ip_1x_1 + ip_2x_2 + ip_3x_3} + \frac{c_{132}}{c_{123}} e^{ip_1x_1 + ip_3x_2 + ip_2x_3} + \right.$$
$$\frac{c_{213}}{c_{123}} e^{ip_2x_1 + ip_1x_2 + ip_3x_3} + \frac{c_{231}}{c_{123}} e^{ip_2x_1 + ip_3x_2 + ip_1x_3} + \frac{c_{312}}{c_{123}} e^{ip_3x_1 + ip_1x_2 + ip_2x_3} \qquad (10.2.12)$$
$$\left. + \frac{c_{321}}{c_{123}} e^{ip_3x_1 + ip_2x_2 + ip_1x_3} \right].$$

We also have

$$\frac{c_{132}}{c_{123}} = S_{32}, \qquad \frac{c_{213}}{c_{123}} = S_{21}, \qquad \frac{c_{231}}{c_{123}} = \frac{c_{231}}{c_{213}} \frac{c_{213}}{c_{123}} = S_{31} S_{21},$$

$$\frac{c_{312}}{c_{123}} = \frac{c_{312}}{c_{132}} \frac{c_{132}}{c_{123}} = S_{31} S_{32}. \qquad \frac{c_{321}}{c_{123}} = \frac{c_{321}}{c_{312}} \frac{c_{312}}{c_{132}} \frac{c_{132}}{c_{123}} = S_{21} S_{31} S_{32}, \qquad (10.2.13)$$

according to the decomposition of any permutation into a sequence of transpositions. If we now set for example $p_2 = p_1 \neq p_3$ we have

$$f_3(x_1, x_2, x_3)|_{p_2=p_1 \neq p_3} = \frac{i^3}{(2\pi)^{\frac{3}{2}}\sqrt{6}} \left[e^{ip_1(x_1 + x_2)+ip_3x_3} + S_{31} e^{ip_1(x_1 + x_3)+ip_3x_2} + \right.$$
$$- e^{ip_1(x_1 + x_2)+ip_3x_3} - S_{31} e^{ip_1(x_1 + x_3)+ip_3x_2} + S_{31}^2 e^{ip_3x_1 + ip_1(x_2 + x_3)} \qquad (10.2.14)$$
$$\left. - S_{31}^2 e^{ip_3x_1 + ip_1(x_2 + x_3)} \right] = 0,$$

which demonstrates the emergence of an exclusion principle in this case as well (bearing again footnote 3 in mind).

This is of course only valid for the continuous spectrum. The discrete spectrum of bound states is more complicated and can be found in [4]. The following exercise shows how bound states work in one particular case, namely the bound state of two particles. In particular, the coupling κ must be negative to have an attractive interaction and the existence of bound states.

Exercise with solution: Two-particle bound state.

Let us recall the two-particle wave function (3.3.17), and let us fix $\kappa = -1$ for simplicity solely for the purposes of this exercise. Analytically continuing the momenta to the complex plane from the 'in' region, we have that, in a neighbourhood of the pole of the function S_{LL},

$$|p_1, p_2\rangle =$$

$$\int_{\mathbb{R}^2} dx_1 dx_2 [\Theta(x_1 - x_2) + S_{LL}(p_1, p_2)\Theta(x_2 - x_1)] e^{ip_1 x_1 + ip_2 x_2} \psi^\dagger(x_1)\psi^\dagger(x_2)|0\rangle \sim$$

$$\frac{-2i}{p_1 - (p_2 - i)} \int_{\mathbb{R}^2} dx_1 dx_2 \Theta(x_2 - x_1) e^{i\frac{p_1 + p_2}{\sqrt{2}}\frac{x_1 + x_2}{\sqrt{2}} + i\frac{p_1 - p_2}{\sqrt{2}}\frac{x_1 - x_2}{\sqrt{2}}} \psi^\dagger(x_1)\psi^\dagger(x_2)|0\rangle \sim$$

$$\frac{-2i}{p_1 - (p_2 - i)} \int_{x_2 > x_1} dx_1 dx_2\, e^{i\frac{p_1 + p_2}{\sqrt{2}}\frac{x_1 + x_2}{\sqrt{2}}} e^{\frac{x_1 - x_2}{2}} \psi^\dagger(x_1)\psi^\dagger(x_2)|0\rangle = (\text{if } x_\pm = \frac{x_1 \pm x_2}{\sqrt{2}})$$

$$\frac{-2i}{p_1 - (p_2 - i)} \int_{-\infty}^{\infty} dx_+\, e^{i\frac{p_1 + p_2}{\sqrt{2}}x_+} \int_{-\infty}^{0} dx_-\, e^{\frac{x_-}{\sqrt{2}}} \psi^\dagger\left(\frac{x_+ + x_-}{\sqrt{2}}\right)\psi^\dagger\left(\frac{x_+ - x_-}{\sqrt{2}}\right)|0\rangle,$$

where we have used $p_1 \sim p_2 - i$ in two of the intermediate stages. As one can notice, the residue (when suitably normalised) describes a normalisable wave function in the relative coordinate x^- thanks to the domain of integration. The state behaves like a plane wave in the center-of-mass coordinate, thereby describing a bound state of two particles whose wave function is peaked where the two are very close to one another (namely, where $|x_-|$ is small, so that the creation operators are creating two particles very near the center of mass). If we appropriately redefine the center-of-mass coordinate to be $X \equiv \frac{x_1 + x_2}{2} = \frac{x_+}{\sqrt{2}}$, then we can see that the plane wave e^{iKX} is such that the momentum of the bound state is simply the sum of the two individual momenta $K = p_1 + p_2$ (which is real—any imaginary part of K would spoil the delta-function normalisability of the center-of-mass part of the wave function).

The energy of the bound state is also easily computed. First, since at the bound state pole the center-of-mass momentum is real and $p_1 = p_2 - i$, we can write

$$p_1 = a - \frac{i}{2}, \qquad p_2 = a + \frac{i}{2}, \qquad a \in \mathbb{R}. \tag{10.2.15}$$

Therefore, we obtain the energy

$$E = p_1^2 + p_2^2 = 2a^2 - 2b^2 = 2a^2 - \frac{1}{2} = \frac{(p_1 + p_2)^2}{2} - \frac{1}{2} \tag{10.2.16}$$

(which is consistently smaller than if the center of mass was freely propagating).

Imagine now the volume to be made finite, namely the theory to be put on an interval $x \in [s_-, s_+] \subset \mathbb{R}$ with periodic boundary conditions. We can follow [3]. In that case, one realises that the spectrum is still organised according to the number of particles present in an energy eigenstate. The wave functions are still given by the same expressions as in the infinite-volume case, the only difference being that *all* the momenta are quantised (even those which used to form the continuous spectrum). The real reason behind the persistence of a form of 'superposition principle' is the underlying integrability of the model, as we will display more clearly further on.

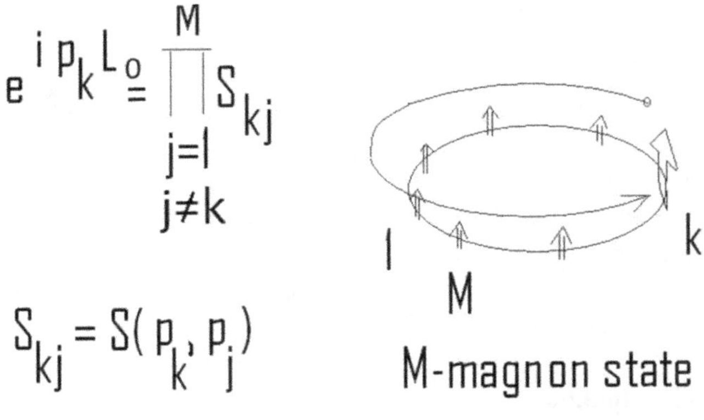

$$e^{i p_k L_0} = \prod_{\substack{j=1 \\ j \neq k}}^{M} S_{kj}$$

$$S_{kj} = S(p_k, p_j)$$

M-magnon state

Figure 10.1. The intuitive picture behind the Bethe equations.

The quantisation condition is encoded in a set of *Bethe equations*:

$$e^{i p_k (s_+ - s_-)} = \prod_{k \neq j = 1}^{M} \frac{p_k - p_j + i\kappa}{p_k - p_j - i\kappa} = \prod_{k \neq j = 1}^{M} S_{LL}(p_k - p_j), \qquad k = 1, \ldots, M. \quad (10.2.17)$$

These are M equations, one for each value of k. The product on the right-hand side runs only over j, from 1 to M, excluding the value $j = k$. While they implement the periodic boundary conditions on the wave functions, the pictorial interpretation of these equations turns out to be rather intuitive. Chosen any one particle k, one can imagine circulating it around the loop. On the one hand, the wave function changes by a phase factor $e^{i p_k L_0}$, where L_0 is the circumference of the loop. On the other hand, the partice scatters against all the others (separately, secretly thanks again to integrability and the fact that scattering is factorised), and so picks up a product of phase factors (S-matrices). The two phase factors need to match (see figure 10.1). The wave function so described is traditionally called the *Bethe wave function*.

One final comment concerns the use of perturbation theory. In [3] it is shown how the use of the Lippmann–Schwinger equation very naturally produces the two-particle Bethe wave function as a result of a resummation of Feynman graphs. One starts from the free wave function

$$|p_1, p_2\rangle_{free} = \int_{\mathbb{R}^2} dx_1 dx_2 \, e^{i k_1 x_1 + i k_2 x_2} \, \psi^\dagger(x_1) \psi^\dagger(x_2) |0\rangle, \quad (10.2.18)$$

and applies the standard Lippmann–Schwinger perturbation-theory argument

$$|p_1, p_2\rangle = \sum_{n=0}^{\infty} \left[G_0(E) \, V \right]^n |p_1, p_2\rangle_{free}, \quad (10.2.19)$$

where

$$G_0(E) = \frac{1}{E - H_0 + i\epsilon}, \quad (10.2.20)$$

$$H_0 = \int_{-\infty}^{\infty} dx \frac{\partial \psi^\dagger}{\partial x} \frac{\partial \psi}{\partial x}, \qquad V = \int_{-\infty}^{\infty} dx \; \psi^{\dagger 2} \psi^2. \qquad (10.2.21)$$

Repeated iteration of this formula produces a series in powers of the coupling constant κ that converges to the expression (3.3.17) (modulo normalisation). Each term can be captured by the drawing of a Feynman diagram with four external legs and a number of bubbles as internal loops, corresponding to the various orders of perturbation theory (see figure 1 in [3]). The S-matrix of this very simple model can then be resummed exactly in terms of a geometric series—precisely as we have discussed in the text below formula (3.3.17).

10.2.2 Integrability method

At this point, we wish to obtain the same result using the integrability approach. This will provide a slightly more geometric derivation, but, most importantly, it is the integrability approach which is then generalised to problems (e.g., Sine-Gordon and Thirring) where the brute-force approach is not available. We start with the monodromy matrix, which is again quantised by normal ordering. This means that we prescribe

$$T(u) \rightarrow :T(u): = \begin{pmatrix} A(u) & \kappa[B(\bar{u})]^\dagger \\ B(u) & [A(\bar{u})]^\dagger \end{pmatrix}, \qquad (10.2.22)$$

with $:.:$ denoting the prescription of rearranging all the $\psi^\dagger$s to the left of all the ψs, and

$$A(u) =: a(u): \qquad \text{and} \qquad B(u) =: b(u): . \qquad (10.2.23)$$

Notice that the trace (transfer matrix) is self-adjoint for real u, and therefore all the conserved charges will be self-adjoint (including of course the Hamiltonian).

The crucial theorem is now the one obtained by Sklyanin. Let us start by considering the theory on a finite interval $x \in [s_-, s_+]$. We will later take the infinite-volume limit to make connections with the results of the previous section. Sklyanin's theorem says that the entries of the monodromy matrix (which are quantum operators) satisfy the RTT relations defined by:

$$R(u - v)T_1(u)T_2(v) = T_2(v)T_1(u)R(u - v). \qquad (10.2.24)$$

Let us unpack these relations a bit further in this particular example. The R-matrix featuring in (10.2.24) is the so-called *Yang's* (rational) R-matrix

$$R(u - v) = 1 \otimes 1 - \frac{i\kappa}{u - v} P, \qquad (10.2.25)$$

where each factor in the tensor product is a $2 \cdot 2$ matrix, and

$$P = \sum_{i,j=1}^{2} E_{ij} \otimes E_{ji} = \begin{pmatrix} 1 & 0 & 0 & 0 \\ 0 & 0 & 1 & 0 \\ 0 & 1 & 0 & 0 \\ 0 & 0 & 0 & 1 \end{pmatrix} \qquad (10.2.26)$$

is the $4 \cdot 4$ permutation matrix. As we have discussed earlier in this book, Yang's R-matrix is known to be the representation of the universal R-matrix of the Yangian associated with $\mathfrak{sl}(2)$ in the fundamental (two-dimensional) representation. The attribute of *rational* is related to the fact that Yangians typically give rise to R-matrices containing rational functions of the spectral parameter. The expression (10.2.25) is the prototype of such R-matrices.

As usual, the notation for the RTT relations means

$$T_1(u) = T(u) \otimes 1, \qquad T_2(v) = 1 \otimes T(v). \tag{10.2.27}$$

The fact that the entries of T are operators on the Fock space of the system means that we should write

$$T(u) = A(u)\, E_{11} + \kappa[B(\bar{u})]^{\dagger}\, E_{12} + B(u)\, E_{21} + [A(\bar{u})]^{\dagger}\, E_{22}, \tag{10.2.28}$$

and keep track of the ordering when we multiply. Let us write everything down explicitly to show how it works:

$$R(u - v)T_1(u)T_2(v) =$$

$$\left[1 \otimes 1 - \frac{i\kappa}{u - v} \sum_{i,j=1}^{2} E_{ij} \otimes E_{ji} \right] \times [A(u)\, E_{11} \otimes 1 + \kappa[B(\bar{u})]^{\dagger}\, E_{12} \otimes 1 + \tag{10.2.29}$$

$$B(u)\, E_{21} \otimes 1 + [A(\bar{u})]^{\dagger}\, E_{22} \otimes 1]$$

$$\times [A(v)\, 1 \otimes E_{11} + \kappa[B(\bar{v})]^{\dagger}\, 1 \otimes E_{12} + B(v)\, 1 \otimes E_{21} + [A(\bar{v})]^{\dagger}\, 1 \otimes E_{22}].$$

We should now multiply the matrices and collect the coefficients in front, bearing in mind that they are non-commutative objects. For instance, one particular term is

$$\left[-\frac{i\kappa}{u - v} \sum_{i,j=1}^{2} E_{ij} \otimes E_{ji} \right] \times [\kappa[B(\bar{u})]^{\dagger}\, E_{12} \otimes 1] \times [[A(\bar{v})]^{\dagger}\, 1 \otimes E_{22}] =$$

$$\frac{-i\kappa^2}{u - v}[B(\bar{u})]^{\dagger}[A(\bar{v})]^{\dagger} \sum_{i,j=1}^{2} E_{ij}E_{12} \otimes E_{ji}E_{22} = \frac{-i\kappa^2}{u - v}[B(\bar{u})]^{\dagger}[A(\bar{v})]^{\dagger}\, E_{22} \otimes E_{12},$$

given that $E_{ab}E_{cd} = \delta_{bc}E_{ad}$. This particular term will have to be collected together with all the other ones proportional to the particular basis element $E_{22} \otimes E_{12}$, and the total coefficient (left-hand side minus right-hand side) will be set to zero. This will give a quadratic relation between the operators $B^{\dagger}$, $A^{\dagger}$, and any other which may be coming from other terms One then has to repeat this procedure by collecting all the coefficients for each separate basis element $E_{ab} \otimes E_{cd}$, for all a, b, c, $d = 1, 2$. To give an example, one of the relations (not actually the one associated with the basis element above) is given by

$$A(u)[B(\bar{v})]^{\dagger} = \left(1 + \frac{i\kappa}{u - v} \right)[B(\bar{v})]^{\dagger}A(u) - \frac{i\kappa}{u - v}[B(\bar{u})]^{\dagger}A(v). \tag{10.2.30}$$

Every one of these $2^4 = 16$ relations is miraculously satisfied by the explicit entries of the monodromy matrix—something that can be painstakingly verified using the fundamental commutation relations (10.2.2). Let us recall in fact that A, B, $A^\dagger$, $B^\dagger$ are very explicit normal-ordered (integral) expressions of the elementary operators ψ, $\psi^\dagger$ obtained from (10.1.6), so in principle we can check each relation using the explicit expressions. The fact that they all work and can be compactly reformulated as RTT relations is one of Sklyanin's most tremendous breakthroughs. Let us first describe how we can use these RTT relations to build the spectrum, and then comment on the algebraic significance of this procedure.

10.2.3 The spectrum via the algebraic Bethe ansatz

We now proceed as follows. We postulate that the true vacuum of the theory is the state $|0\rangle$, which is annihilated by $\psi(x)$ for all real values of x. By inspection of the form of $B(u)$ (10.1.6), the latter being already in normal-ordered form upon $\psi^* \to \psi^\dagger$, we see a systematic excess of ψs over $\psi^\dagger$ s. Therefore, we can certainly deduce that

$$B(u)|0\rangle = 0. \tag{10.2.31}$$

Remark: Notice that in the literature on the Bethe ansatz C typically annihilates and B creates. We maintain here the historical notation of [4], where B annihilates and $B^\dagger$ creates. We hope that this choice will not confuse the reader.

A relation of the type (10.2.30) suggests that $B^\dagger$ might then create exact eigenstates because it seems to have the appropriate kind of behaviour when being near an A (the latter being a part of the transfer matrix). In other words, we postulate that the exact spectrum is built as follows:

$$|g_k\rangle \equiv B^\dagger(p_1)...B^\dagger(p_M)|0\rangle, \tag{10.2.32}$$

where all the momenta p_i are real. This means that we shall describe in this fashion the continuous spectrum. Bound states in the attractive $\kappa < 0$ case will be obtained by studying the complex solutions to the Bethe equations obtained at the end of this procedure.

The order of the $B^\dagger$s in (10.2.32) does not matter because another one of the relations descending from RTT is that two $B^\dagger$s commute for different values of the respective arguments:

$$B^\dagger(p_i)B^\dagger(p_j) = B^\dagger(p_j)B^\dagger(p_i), \qquad p_i, p_j \in \mathbb{R}. \tag{10.2.33}$$

If we now act on (10.2.32) with the transfer matrix, then we can use the relationships descending from RTT to commute the operators through: if we take u to be real as well,

$$[A(u) + A^\dagger(u)]B^\dagger(p_1)...B^\dagger(p_M)|0\rangle =$$

$$\left(1 + \frac{i\kappa}{u - p_1}\right)B^\dagger(p_1)A(u)B^\dagger(p_2)...B^\dagger(p_M)|0\rangle - \frac{i\kappa}{u - p_1}B^\dagger(u)A(p_1)B^\dagger(p_2)...B^\dagger(p_M)|0\rangle + \cdots$$

One has to repeatedly use the RTT relations until both A and $A^\dagger$ have gone through all the $B^\dagger$ s, and finally one can simply rely on the fact that the vacuum is an exact eigenstate of A and $A^\dagger$ by explicit computation from (10.1.6) (where all the expressions are already normal-ordered upon $\psi^* \to \psi^\dagger$):

$$[A(u) + A^\dagger(u)]|0\rangle = 2\cos\frac{u}{2}(s_+ - s_-)|0\rangle. \tag{10.2.34}$$

Collecting all the terms, one can easily see that there is one single contribution out of the many that actually reproduces the initial state (see the first term on the right-hand side of (10.2.3), and imagine that term iterated M times), and therefore that contribution provides the eigenvalue:

$$[A(u) + A^\dagger(u)]B^\dagger(p_1)...B^\dagger(p_M)|0\rangle = \Lambda(u; p_1, \ldots, p_M)B^\dagger(p_1)...B^\dagger(p_M)|0\rangle + \cdots \tag{10.2.35}$$

(we shall not need here the explicit form of the eigenvalue Λ). There are also a lot of other terms, which, however, do not reproduce the initial states—some are coming for instance from iterations of the first term on the right-hand side of (10.2.3). We need to make sure that these extra *unwanted* terms do cancel, and that is the task of the *Bethe equations*. Imposing the system of Bethe equations

$$e^{ip_i(s_+ - s_-)} = \prod_{i \neq j=1}^{M} S_{ij}, \tag{10.2.36}$$

can be seen to cancel all the unwanted terms, thereby leaving us with an eigenstate. The functions S_{ij} exactly coincide with the function (10.2.10) (with the replacement $p_1 \to p_i$ and $p_2 \to p_j$).

The system of equations (10.2.36) can be interpreted as imposing a quantisation condition on the momenta $p_1, \ldots, p_M$. If the theory were free, namely $\kappa = 0$ hence $S_{ij} = 1$, then we would in fact obtain from (10.2.36)

$$p_i = \frac{2\pi n_i}{(s_+ - s_-)}, \qquad n_i \in \mathbb{Z}, \qquad i = 1, \ldots, M. \tag{10.2.37}$$

In infinite volume the relations are slightly different, which produces the result that the unwanted term is automatically zero. This fits with the fact that there is no quantisation condition for the momenta on the infinite line.

It is now not too surprising, but striking nevertheless, that the explicit form of the wave functions reproduces exactly (up to normalisation) the brute-force diagonalisation result that we displayed earlier [3]:

$$\langle x_1, \ldots, x_M | B^\dagger(p_1)...B^\dagger(p_M)|0\rangle \sim \langle 0|\psi(x_1)...\psi(x_M)|B^\dagger(p_1)...B^\dagger(p_M)|0\rangle \propto f_M(x_1, \ldots, x_M).$$

10.3 The classical limit

One of the advantages of the integrability approach is the possibility of reducing the quantum to classical transition to a purely algebraic (quantum group) problem, as we have alluded to earlier. In fact, let us consider the fundamental RTT relations (10.2.24)

$$R(u - v)T_1(u)T_2(v) = T_2(v)T_1(u)R(u - v). \tag{10.3.1}$$

By plugging into (10.3.1) the explicit form of the R-matrix (10.2.25)

$$R(u - v) = 1 \otimes 1 - \frac{i\kappa}{u - v}P \equiv 1 \otimes 1 + i\hbar\, r(u - v), \tag{10.3.2}$$

where $\hbar$ has been reinstated to guide our physical intuition and the coupling constant also appears, and as a consequence $r(u - v) \equiv -\kappa\frac{P}{u-v}$, one obtains

$$r(u - v)T_1(u)T_2(v) - T_2(v)T_1(u)r(u - v) = -\frac{1}{i\hbar}[T_1(u), T_2(v)]. \tag{10.3.3}$$

Imagine now sending $\hbar \to 0$. In this limit the operators tend to classical commuting variables, therefore the left-hand side of (10.3.3) tends to $[r(u - v), T_1(u)T_2(v)]$. The right-hand side has a $\frac{0}{0}$ because the commutator of two classical variables would be zero. This undetermined form tends to the classical Poisson brackets because $\{.,.\} \sim \frac{[.,.]}{-i\hbar}$. In the limit we therefore get

$$\{T_1(u), T_2(v)\} = [r(u - v), T_1(u)T_2(v)]. \tag{10.3.4}$$

Lo and behold, this is exactly the classical Poisson structure of the Lieb–Liniger model—see [2] formula (3.35). In fact, P coincides with the quadratic Casimir of the tensor algebra $\mathfrak{su}(2) \otimes \mathfrak{su}(2)$. Moreover, the object $r(u - v)$ is precisely the classical r-matrix—formula (4.16) in [2], the solution of the so-called *classical* Yang–Baxter equation.

We therefore have an algebraic way of connecting classical and quantum physics, very concretely in this specific physical encarnation: the integrable structures are simply related by the transition between classical r- versus quantum R-matrices, which is completely under control from the Hopf-algebra point of view (see also the discussion around (4.13)–(4.15) in [2]). Although in this very special case we have that $R = 1_4 + i\hbar r$, in general the $\hbar$ series does not truncate, and we can only write

$$r(u - v) = -i\frac{\partial}{\partial\hbar}R(u - v)_{|\hbar=0}. \tag{10.3.5}$$

It is also very interesting to note[1] that the effect of the normal ordering in this particular problem is effectively to produce a constant shift

$$\begin{array}{ccc} r(u - v) \to & [\text{normal ordering}] & \to R(u - v) = 1_4 + i\hbar\, r(u - v) \\ T(u) & : T(u): & \end{array} \tag{10.3.6}$$

(very similar to the zero-point shift of a harmonic oscillator). The appearance of the constant shift due to the normal ordering can actually be explicitly traced in the calculation [4].

---

[1] We thank Evgeny Sklyanin for private communication on this point.

References

[1] Avan J, Caudrelier V, Doikou A and Kundu A 2016 Lagrangian and Hamiltonian structures in an integrable hierarchy and space-time duality *Nucl. Phys.* B **902** 415–39

[2] Torrielli A 2016 Lectures on classical integrability *J. Phys.* A **49** 323001

[3] Thacker H B 1981 Exact integrability in quantum field theory and statistical systems *Rev. Mod. Phys.* **53** 253

[4] Sklyanin E K 1980 Quantum version of the method of inverse scattering problem *Zap. Nauchn. Semin.* **95** 55–128

[5] Klose T and Zarembo K 2006 Bethe ansatz in stringy sigma models *J. Stat. Mech.* **0605** P05006

IOP Publishing

Integrability using the Sine-Gordon and Thirring Duality
An introductory course
Alessandro Torrielli

Chapter 11

Supplement: massless integrability

In this part of the supplement, we review the theory of massless integrable scattering. This is not (just) the limit to zero mass of a massive scattering theory. It is a much richer subject, which has to do with the trajectories to and from conformal fixed points of renormalisation group flows, and therefore it constitutes a crucial link between $1 + 1$-dimensional integrable quantum field theories and $1 + 1$-dimensional conformal field theories (CFTs). In this part of the supplement we primarily follow [1, 2]—see also [3–5].

11.1 The limit to zero mass

It is instructive to perform the limit $m \to 0$ starting from a massive integrable scattering theory. This means that we shall assume that we do have an exact integrable relativistic S-matrix $S(\theta_1 - \theta_2)$ to begin with, and that both scattering particles have the same mass $m > 0$. The dispersion relation is

$$E_i = m \cosh \theta_i, \quad p_i = m \sinh \theta_i. \tag{11.1.1}$$

To take the massless limit we simultaneously boost the rapidities to $\pm\infty$, which means that the corresponding velocities tend to the speed of light ± 1 (in our units), respectively. In one spatial dimension it makes sense to decide whether a massless particle travels to the right or to the left: in fact, once we have taken the limit in one or the other way, there will be no Lorentz transformation that can change the fact that the particle is moving right, or left, with the speed of light.

In formulas, we have for a right mover

$$m \to 0, \quad \theta_i \to \Theta + \vartheta_i, \quad \Theta \to +\infty, \tag{11.1.2}$$

while for a left mover we have

$$m \to 0, \quad \theta_i \to -\Theta + \vartheta_i, \quad \Theta \to +\infty. \tag{11.1.3}$$

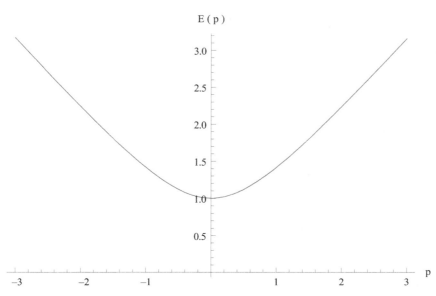

Figure 11.1. The massive dispersion relation $E_i^2 = p_i^2 + m^2$. The two arms of the graph are connected by a Lorentz boost.

The variables ϑ_i are kept finite in the limit, whereas

$$\frac{m}{2}e^{\Theta} \to \mu < \infty \qquad (11.1.4)$$

In this fashion we achieve the massless dispersion relation: for a right mover

$$E_i = \frac{m}{2}(e^{\theta_i} + e^{-\theta_i}) \to \mu e^{\vartheta_i}, \quad p_i = \frac{m}{2}(e^{\theta_i} - e^{-\theta_i}) \to \mu e^{\vartheta_i}, \quad E_i \to p_i, \qquad (11.1.5)$$

while for a left mover we have

$$E_i = \frac{m}{2}(e^{\theta_i} + e^{-\theta_i}) \to \mu e^{-\vartheta_i}, \quad p_i = \frac{m}{2}(e^{\theta_i} - e^{-\theta_i}) \to -\mu e^{-\vartheta_i}, \quad E_i \to -p_i. \quad (11.1.6)$$

These are indeed the two branches of the massless dispersion relation, which splits into two at a cusp—see figures 11.1 and 11.2. It is quite clear that there are now four possible limits which one can take on the S-matrix because either particle can be sent off to be a right or a left mover. Let us divide them into two sets: 'same' and 'opposite'.

If the two particles are taken in the 'same' limit, meaning both right or both left movers, then we expect that the scattering will be nonperturbative in nature, meaning that there will be no regime where the scattering matrix will smoothly reduce to the permutation matrix (analogous to the R-matrix reducing to the identity). In these two situations, it is not immediately clear what the physical meaning of these S-matrices will be, but we shall shortly discover it by following Zamolodchikov.

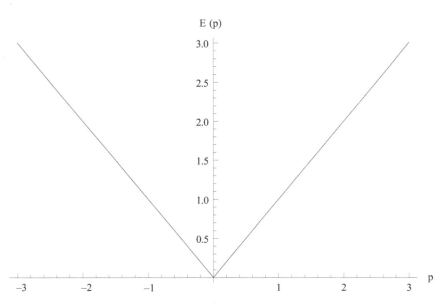

Figure 11.2. The massless dispersion relation $E_i = |p_i^2|$. The two arms of the graph are no longer connected by a Lorentz boost, and are parameterised by $E_i = \pm p_i = \pm \mu e^{\pm \vartheta_i}$, respectively, with $\vartheta_i \in (-\infty, \infty)$ separately in either case.

If the two particles are taken in two 'opposite' limits, meaning right-left or left-right, then we expect the existence of a regime where the scattering is trivial (the S-matrix smoothly reducing to the permutation). Nevertheless, despite this situation looking more normal, the nonexistence of a mass gap jeopardises in principle a lot of the properties of integrable scattering. Once again following Zamolodchikov, we will shortly see how to interpret this occurrence.

We shall also remark that having dealt with the exact S-matrices from the start, we do not incur the notorious issues presented by the Feynman-integral perturbation theory of a massless particle that could be a scalar. The perturbative expansion of the right-left and left-right S-matrices can be performed by simple Taylor series of the exact expressions, without any reference to Feynman integrals[1].

11.2 Massless flows

To help us understand what to do in this situation, Zamolodchikov observes that the four possible limits of the S-matrix (right-right, left-left, right-left and left-right) are intrinsically very different.

Let us start from the opposite scattering, namely right-left and left-right. Because the S-matrix only depends on the difference of the rapidities, we will have

$$S(\theta_1 - \theta_2) = S(\pm 2\Theta + \vartheta_1 - \vartheta_2), \tag{11.2.1}$$

---

[1] We thank Ctirad Klimčík for a discussion about this point.

where $\pm 2\Theta$ corresponds to right-left or left-right, respectively. The large offset Θ either way does not cancel, and it is then sent to infinity. The S-matrix is often made of ratios of functions, and when the argument becomes very large in modulus, numerator and denominator often become the same (up to signs), and the S-matrix often tends to trivialise—it often becomes the permutation. It is hard to make this statement very precise, in fact it does not really have to be true in all cases—but it often does. One might approach this point by arguing that the S-matrix often tends to trivialise when one of the two scattering particles has got infinitely more rapidity than the other one—as happens for a right mover versus a left mover, the rapidity difference being $+\infty - (-\infty)$. When this happens, the left and right modes decouple, they become transparent to each other and cease to interact—the mutual S-matrix is trivial. This is a sign of a CFT, where right and left modes are decoupled.

The right-right and left-left situations are quite different. In this case we have

$$S(\theta_1 - \theta_2) = S(\vartheta_1 - \vartheta_2), \tag{11.2.2}$$

the large offset Θ having cancelled out. As we have seen in Sine-Gordon and as it is often the case, the mass does not really feature directly in the S-matrix entries when expressed in the rapidity variable. This means that effectively nothing changes for the right-right and left-left scattering—at least formally, meaning that the functional form of the S-matrix remains as it was in the massive case (with a different dispersion relation for the particles though). Zamolodchikov argues that these two S-matrices (right-right being a purely 'right' object and left-left being a purely 'left' object) characterise the CFT that one obtains in the limit (having ascertained earlier that left and right modes often do not interact in the massless limit). It is fruitful to consider these same-limit matrices as a tool for generating observables in the right and left sector of the limiting CFT.

It is possible to reinforce the argument by showing that the resulting theory has now effectively lost all its scale dependence. In fact, if we focus on the right-right case for example, we have that $S(\vartheta_1 - \vartheta_2)$ does not have any mass scale in it, but we still have

$$E_i = p_i = \mu \, e_i^\vartheta, \quad i = 1, 2. \tag{11.2.3}$$

However, we can write

$$E_i = p_i = e^{\tilde\vartheta_i}, \quad \tilde\vartheta_i = \vartheta_i + \log \mu, \quad i = 1, 2, \tag{11.2.4}$$

and

$$S(\vartheta_1 - \vartheta_2) = S(\tilde\vartheta_1 - \tilde\vartheta_2). \tag{11.2.5}$$

The S-matrix, which characterises the right sector, is completely insensitive to the mass scale (similarly for the left sector). This perfectly embodies the idea of a theory being at a critical point: dimensional analysis forces the theory to contain parameters that have the dimension of a mass—however, rescaling those parameters

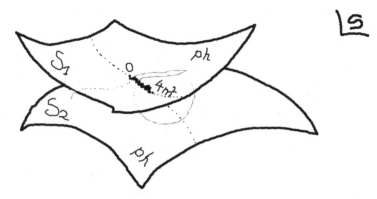

Figure 11.3. The two Riemann sheets of the S-matrix in the case of quadratic branch cuts.

('zooming in' or 'out') leaves the theory, when re-expressed in the rescaled variables, exactly unchanged[2]. We see this happening precisely in the right-right and left-left massless scattering matrices, which control the dynamical content of the critical point.

The idea of characterising a CFT as a particular integrable quantum field theory is the subject of an important series of works [6–9], see also [10]. As we have motivated in this book, $c = 1$ CFTs are particularly relevant for the ultraviolet limit of Sine-Gordon/Thirring.

There is, however, a question that arises at this point: what we have outlined seems like an interesting way to describe a CFT, but it is a bit too abrupt. In particular, it would be desirable to be able to reinstate the left-right interaction in a controlled way, and to be able to move away from the fixed point while keeping the theory massless. In other words, can we have a situation where the theory is massless but not scale-invariant?

The answer is yes, in that Zamolodchikov has shown how to construct a nontrivial right-left interaction for massless modes [2]. Since there is no mass gap, one has to invent a special set of 'axioms' which fit the purpose. Let us display the procedure for a particular class of models for simplicity, namely models with quadratic branch cuts. This means that the branch cuts in figure 2.5 are of quadratic type, and one comes back to the original sheet after going twice around the branch points. This is of course not the case of the Sine-Gordon S-matrix, which has got infinitely many Riemann sheets and is slightly harder to play with as an example.

Because of the quadratic branch cuts, we can draw a picture such as figure 11.3. We have now changed the choice of the branch cut and drawn it on the real axis *between*0 and $4m^2$ (making the choice of a *short cut* as opposed to a *long cut*). As a consequence, the *physical sheet* is now split into two regions:

---

[2] We thank Ingo Runkel and Patrick Dorey for pointing this out to us.

Superior upper half plane, where $S(s) = S_1(s)$ and Im $s > 0$,

Inferior lower half plane, where $S(s) = S_2(s)$ and Im $s < 0$,

with $S_1(s)$ and $S_2(s)$ being the value of the function on the superior, resp. inferior sheet in figure 11.3. Figure 11.3 also reports the two paths involved in braiding unitarity and crossing. The path from s in the superior part of the physical sheet, to the same value s but in the unphysical part of the inferior sheet, is related to braiding unitarity:

$$S_1(s)S_2(s) = 1, \tag{11.2.6}$$

(assuming, again for simplicity, no internal degrees of freedom and the S-matrix being just a scalar factor). The path from s in the superior part of the physical sheet, to $4m^2 - s$ in the physical part of the inferior sheet, is related to crossing:

$$S_1(s) = S_2(4m^2 - s). \tag{11.2.7}$$

It is now relatively straightforward to take the $m \to 0$ limit. First, the branch cut closes up and one ends up with two almost disconnected sheets, touching at a point (the origin). Second, the braiding unitarity and crossing relations reduce to

$$S_1(s) = S_2(-s), \quad S_i(s)S_i(-s) = 1, \quad i = 1, 2. \tag{11.2.8}$$

There is no longer any room for bound state poles because the pair-production threshold has gone to 0, which fits with the idea that two massless particles cannot form a bound state. The combination of (11.2.8), the S-matrix being a phase for real values of the rapidities, and the statement of the absence of bound states, makes for a significant part of the new 'axioms' of massless right-left scattering.

The left-right scattering can be understood as the time-reversed of the right-left scattering, and therefore it involves the inverse of the S-matrix.

It is important to point out that the relevance of the value of the limiting S-matrix for $\vartheta_1 = \vartheta_2$ in ascertaining the statistical behaviour of particles, in the way which we have often previously elaborated upon, is lost in the massless case for the right-left and left-right scattering. For example, the condition of braiding unitarity becomes the uncommunicative $S_i(4\mu^2)S_i(-4\mu^2) = 1$ (using that $s = 4\mu^2 e^{\vartheta_1 - \vartheta_2}$). We can think of this as effectively being due to the two distinct infinite boosts, which make the right- and left-particle rapidities conceptually incomparable. In the massless case, left and right movers are to be thought of as genuinely different particles.

11.2.1 Tricritical to critical Ising

The canonical example for the new 'axioms' that we have just discussed is the renormalisation group flow between the following two different CFTs: the tricritical point of the Ising model (as the ultraviolet fixed point) and the critical point of the same model (as the infrared fixed point) [2]. The idea of such a massless flow is that one can begin by first considering the infrared fixed point, where the theory is actually very simple: it is described by a free *Majorana* fermion with central charge $c = \frac{1}{2}$. The scattering particle has one single degree of freedom (the S-matrix is a scalar factor) and the right-left S-matrix is trivial at the infrared fixed point. The CFT being *a free CFT* implies a further simplification with respect to what we discussed: also the right-right and the left-left S-matrices are trivial at the fixed point. They would not be trivial if the CFT were interacting. Moreover, since the functional form of the right-right and left-left S-matrices is scale-independent, as we have seen, we conclude that the right-right and left-left S-matrices are trivial along the entire massless flow.

The right-left S-matrix for this flow has been found by Zamolodchikov to be [2]

$$S_1(s) = \frac{4i\mu^2 - s}{4i\mu^2 + s}, \quad S_2(s) = S_1(-s) = \frac{1}{S_1(s)}. \tag{11.2.9}$$

This S-matrix satisfies all the axioms of massless right-left scattering, and it is also *real analytic*: $S_1(s^*) = S_2^*(s)$, given that $\mu \in \mathbb{R}$. The dispersion relation naturally reads

$$E_1 = p_1 = \mu e^{\vartheta_1}, \quad E_2 = -p_2 = \mu e^{-\vartheta_2}, \quad s = 4\mu^2 e^{\vartheta_1 - \vartheta_2}. \tag{11.2.10}$$

We have therefore

$$S_1(\vartheta_1 - \vartheta_2) = \frac{i - e^{\vartheta_1 - \vartheta_2}}{i + e^{\vartheta_1 - \vartheta_2}} = -\tanh\left(\frac{\vartheta_1 - \vartheta_2}{2} - i\frac{\pi}{4}\right), \quad S_2(\vartheta_1 - \vartheta_2) = \frac{1}{S_1(\vartheta_1 - \vartheta_2)}$$

and

$$S_1(\vartheta_1 - \vartheta_2) = \frac{1}{S_1(\vartheta_1 - \vartheta_2 + i\pi)} \quad \text{hence} \quad S_i(\vartheta_1 - \vartheta_2)S_i(\vartheta_1 - \vartheta_2 + i\pi) = 1. \tag{11.2.11}$$

Notice that there are no bound states. The simple pole of $S_1(s)$ is located at $-4i\mu^2$, which is in the superior lower half plane in figure 11.3 (away from the physical region—this pole actually signifies the presence a so-called *resonance*). We should also remark that the antiparticle of a right mover is a right mover, and the antiparticle of a left mover is a left mover, according to the crossing transformation $\vartheta_i \to \vartheta_i + i\pi$. Such transformation changes sign simultaneously to the energy and the momentum of an individual particle, leaving the dispersion relation unchanged.

To demonstrate the validity of this S-matrix, Zamolodchikov derives the thermodynamic Bethe ansatz (TBA) from this S-matrix and computes the effective central charge, which we have argued in the main text is related to the ground state energy produced by the TBA. The result is precisely $c = \frac{1}{2}$ in the infrared, and $c = \frac{7}{10}$ in the ultraviolet, corresponding to the two fixed points of the tricritical to critical Ising model (and in accordance with the decreasing monotonicity of the effective central charge from the ultraviolet to the infrared).

Plenty of other examples of massless flows exist, in particular the flow between minimal model CFTs. We refer to the review [1] for an excellent starting point to explore the literature on this topic, which is once again extremely vast.

For a review of conformal field theories the reader can consult for example [11–13].

11.3 Thermodynamic Bethe ansatz for a simple S-matrix

We briefly review here how the TBA [14]—see [15] for an excellent review—works for the case of the tricritical to critical Ising flow that we have just discussed. Since the S-matrix is very simple and there are no bound states, this is an ideal pedagogical starting point to learn the basic idea. The TBA for the Sine-Gordon model is naturally much more complicated, and we refer to the literature on the subject for the full detail—see for instance [16–18].

The first thing to remember from the discussion that we have just gone through is that the scattering involves two types of particles: right movers with momentum $p_1 = \mu e^{\vartheta}$ and left movers with momentum $p_2 = -\mu e^{-\vartheta_2}$. The scattering matrix is given by

$$S_1(\vartheta_1 - \vartheta_2) = -\tanh\left(\frac{\vartheta_1 - \vartheta_2}{2} - i\frac{\pi}{4}\right) \equiv S(\vartheta) \equiv S(p_1, p_2), \quad \vartheta \equiv \vartheta_1 - \vartheta_2,$$

in the conventions of the previous discussion. We shall simply denote by S the only S-matrix relevant to this discussion. Let us decide for simplicity to denote with the letter p the right movers, and with the letter q the left movers.

We can write the Bethe equations which quantise the momenta of M_R right movers and M_L left movers on a circle of length L (this is a circle in the doubly-Wick-rotated theory [14]):

$$e^{ip_iL} \prod_{j\neq i}^{M_R} S(p_i, q_j) = 1, \quad e^{iq_kL} \prod_{j\neq k}^{M_L} S(q_k, p_j) = 1, \quad i = 1, \ldots, M_R, \quad k = 1, \ldots, M_L,$$

where the first equation quantises the momenta of the right movers (as resulting from the interaction with the left movers), and the second equation quantises the momenta of the left movers (as resulting from the interaction with the right movers).

We now take the logarithm of both sides of both equations, and divide by $2\pi i$:

$$\frac{Lp_i}{2\pi} - \frac{i}{2\pi}\sum_{j\neq i}^{M_R} \log S(p_i, q_j) = n_i, \quad n_i \in \mathbb{Z},$$

$$\frac{Lq_k}{2\pi} - \frac{i}{2\pi}\sum_{j\neq i}^{M_L} \log S(q_k, p_j) = m_k, \quad m_k \in \mathbb{Z}.$$

(11.3.1)

We then take the *thermodynamic limit*: this involves sending M_R, M_L and L to infinity. Given that free particles have momenta which are quantised as integer multiples of $\frac{2\pi}{L}$, we can expect the spacing between the quantised levels to become finer and finer in the thermodynamic limit. We can describe the situation with two types of densities: $\rho_i^p(\vartheta)$ and $\rho_i^h(\vartheta)$, $i = 1, 2$. The index $i = 1$ is for right movers, while $i = 2$ is for left movers. The density ρ^p is for *particles*, meaning the quantised levels which are occupied, and ρ^h is for *holes*, meaning the quantised levels which are not occupied. We also define $\rho = \rho^p + \rho^h$. We turn the sum into an integral and write down the thermodynamic form of the Bethe equations:

$$\frac{\mu L\, e^{\vartheta}}{2\pi} + \frac{1}{2\pi}\int_{-\infty}^{\infty} d\beta\; \varphi(\vartheta - \beta)\rho_2^p(\beta) = \rho_1^p(\vartheta) + \rho_1^h(\vartheta) = \rho_1(\vartheta),$$

$$\frac{\mu L\, e^{-\vartheta'}}{2\pi} + \frac{1}{2\pi}\int_{-\infty}^{\infty} d\beta\; \varphi(\vartheta' - \beta)\rho_1^p(\beta) = \rho_2^p(\vartheta') + \rho_2^h(\vartheta') = \rho_2(\vartheta') \tag{11.3.2}$$

(for a step-by-step derivation, involving a process of finite differences turning into continuous differentials, the reader is referred, for instance, to section 2.1.2 of [19]). In reaching this form we have remembered the expression of the momentum in terms of the rapidity for right and left movers, and used the fact that the index becomes a continuum variable in the thermodynamic limit: to keep the notation distinct, we use $p_i \to \mu e^{\vartheta}$ and $q_k \to -\mu e^{-\vartheta'}$ in the thermodynamic limit. We have also introduced the kernel associated with the S-matrix:

$$\varphi(x) = -i\frac{d}{dx}\log S(x) = \frac{1}{\cosh x}. \tag{11.3.3}$$

For future reference, we can take the variation of the integral form of the Bethe equations when the densities vary:

$$\frac{1}{2\pi}\varphi * \delta\rho_2^p = \delta\rho_1, \quad \frac{1}{2\pi}\varphi * \delta\rho_1^p = \delta\rho_2, \tag{11.3.4}$$

if $*$ denotes the convolution, defined as $a*b = \int_{-\infty}^{\infty} dy\, a(x - y)\, b(y)$.

The essence of the TBA comes at this point. The idea, which the reader can find very clearly explained in the original [14], is that the ground state energy of the system on a circle of size L_0 (this being the circle in the original theory) is equivalently obtained by minimising the free energy for the doubly-Wick-rotated system. In the analysis, the relativistic invariance of the model is exploited. One is therefore led to the problem of minimising the free energy functional $F = H - TS$ for variations of the densities subject to the constraints (11.3.4). The pieces assembling the free energy are the total energy H of the double-Wick-rotated theory (which gets contributions from the occupied levels)

$$H = \sum_{i=1}^{2}\int_{-\infty}^{\infty} d\beta\; \mu\, e^{\sigma_i\beta}\, \rho_i^p(\beta), \tag{11.3.5}$$

where $\sigma_1 = +$ and $\sigma_2 = -$, and the related entropy, given in statistical form as

$$S = \sum_{i=1}^{2}\int_{-\infty}^{\infty} d\beta\Big[\rho_i \log \rho_i - \rho_i^p \log \rho_i^p - \rho_i^h \log \rho_i^h\Big]. \tag{11.3.6}$$

At this point it is worth specifying explicitly that the scattering particles are fermions. We did not need to state it explicitly before, but the entropy (11.3.6) is the appropriate one for fermionic particles [14]. The corresponding expression for bosons is slightly different, and produces slightly different formulas at the end of the TBA procedure. The correct evaluation of the central charge depends on the appropriate implementation of the entropy formula.

We also have that the temperature T in the doubly-Wick-rotated picture is given by $\frac{1}{L_0}$. If we now vary the free energy, and use the fact that

$$\delta(x \log x) = (1 + \log x)\delta x, \qquad (11.3.7)$$

we obtain

$$\delta F = \sum_{i=1}^{2} \int_{-\infty}^{\infty} d\beta \left[\mu \, e^{\sigma_i \beta} \delta \rho_i^p - \frac{1}{L_0} \log \rho_i \, \delta \rho_i + \frac{1}{L_0} \log \rho_i^p \, \delta \rho_i^p + \frac{1}{L_0} \log \rho_i^h \, \delta \rho_i^h \right],$$

where we have cancelled the 1s coming from (11.3.7) since $\rho_i = \rho_i^p + \rho_i^h$. Straightforward manipulations bring this to

$$\delta F = \sum_{i=1}^{2} \int_{-\infty}^{\infty} d\beta \left[\mu \, e^{\sigma_i \beta} \delta \rho_i^p - \frac{1}{L_0} \log \frac{\rho_i^h}{\rho_i^p} \, \delta \rho_i^p - \frac{1}{L_0} \log(1 + \frac{\rho_i^p}{\rho_i^h}) \, \delta \rho_i \right].$$

We now use (11.3.4), and rewrite the resulting terms $\int \log(1 + \frac{\rho_1^p}{\rho_1^h}) \, \varphi * \delta \rho_2^p$ and $\int \log(1 + \frac{\rho_2^p}{\rho_2^h}) \, \varphi * \delta \rho_1^p$ as $\int \delta \rho_2^p \, \varphi * \log(1 + \frac{\rho_1^p}{\rho_1^h})$ and $\int \delta \rho_1^p \, \varphi * \log(1 + \frac{\rho_2^p}{\rho_2^h})$, respectively. We can do this because the kernel is an even function of its argument. This means that we can now collect all the terms that multiply $\delta \rho_i^p$, and set them separately to zero, for $i = 1$ and $i = 2$, since the variations of the two densities for right and left movers are independent. We obtain in this way two separate equations:

$$-\mu L_0 \, e^{\vartheta} + \log \frac{\rho_1^h}{\rho_1^p} + \frac{1}{2\pi} \varphi * \log(1 + \frac{\rho_2^p}{\rho_2^h}) = 0,$$

$$-\mu L_0 \, e^{-\vartheta'} + \log \frac{\rho_2^h}{\rho_2^p} + \frac{1}{2\pi} \varphi * \log(1 + \frac{\rho_1^p}{\rho_1^h}) = 0. \qquad (11.3.8)$$

At last, defining the so-called *pseudoenergies*

$$\varepsilon_i = \log \frac{\rho_i^h}{\rho_i^p}, \quad i = 1, 2, \qquad (11.3.9)$$

we obtain the TBA equations

$$-\mu L_0 \, e^{\vartheta} + \varepsilon_1 + \frac{1}{2\pi} \varphi * \log(1 + e^{-\varepsilon_2}) = 0,$$

$$-\mu L_0 \, e^{-\vartheta'} + \varepsilon_2 + \frac{1}{2\pi} \varphi * \log(1 + e^{-\varepsilon_1}) = 0. \qquad (11.3.10)$$

Let us point out that we have reached such a nice and compact form for the TBA because there are no bound states. If there were bound states, then they would come with their own additional pseudoenergies, which would bring in additional associated equations.

We can now study the ultraviolet and infrared regimes and extract the respective central charges. Let us begin with the infrared first. The infrared regime corresponds to $\mu L_0 > > 1$, hence from (11.3.10) we get that

$$\varepsilon_1 \sim \mu L_0 e^{\vartheta}, \quad \varepsilon_2 \sim \mu L_0 e^{-\vartheta'}, \tag{11.3.11}$$

is an asymptotic solution of the TBA equations (the kernel part consistently going to zero). We need to use the fact that the minimum total free energy in the doubly-Wick-rotated theory, which in turns provides the value of the ground state energy of the original theory, is given by

$$E_{min} = \frac{L_0}{L} F_{min} = -\frac{\mu}{2\pi} \sum_{i=1,2} \int_{-\infty}^{\infty} d\beta \, e^{\sigma_i \beta} \log(1 + e^{-\varepsilon_i}).$$

A sketch of the proof that the minimum of F is given as in (11.3) is provided, for instance, in formula (2.42) of [19]. The proof, which we will reproduce now, uses both the TBA and the integral form of the Bethe equations in addition to the even nature of the kernel to be able to shift the convolution. The proof shows in particular that the value at the minimum only depends on the ratios of the type $\frac{\rho^p}{\rho^h}$. Here is the complete write-out for our situation: first, let us define

$$\bar{1} \equiv 2, \quad \bar{2} \equiv 1, \quad \int \equiv \int_{-\infty}^{\infty}. \tag{11.3.12}$$

We then have

$$E = \frac{1}{L}\sum_{i=1}^{2} \int d\beta \Big[\mu L_0 e^{\sigma_i \beta} - (\rho_i^p + \rho_i^h)\log(\rho_i^p + \rho_i^h) + \rho_i^p \log \rho_i^p + \rho_i^h \log \rho_i^h \Big] =$$

$$\frac{1}{L}\sum_{i=1}^{2} \int d\beta \left[\rho_i^p \left(\mu L_0 e^{\sigma_i \beta} + \log \frac{\rho_i^p}{\rho_i^p + \rho_i^h} \right) + \rho_i^h \log \frac{\rho_i^h}{\rho_i^p + \rho_i^h} \right] = (\text{using the TBA equations}) \tag{11.3.13}$$

$$\frac{1}{L}\sum_{i=1}^{2} \int d\beta \left[\rho_i^p \left(\log \frac{\rho_i^h}{\rho_i^p} + \frac{1}{2\pi} \varphi * \log \frac{\rho_i^p + \rho_i^h}{\rho_i^h} + \log \frac{\rho_i^p}{\rho_i^p + \rho_i^h} \right) + \rho_i^h \log \frac{\rho_i^h}{\rho_i^p + \rho_i^h} \right].$$

Now we use the fact that the kernel is even to shift the convolution, and we also combine two pieces:

$$E = \frac{1}{L}\sum_{i=1}^{2} \int d\beta \left[\log \frac{\rho_i^p + \rho_i^h}{\rho_i^h} \frac{1}{2\pi} \varphi * \rho_i^p + \rho_i^p \log \frac{\rho_i^h}{\rho_i^p + \rho_i^h} + \rho_i^h \log \frac{\rho_i^h}{\rho_i^p + \rho_i^h} \right]. \tag{11.3.14}$$

Now we use the Bethe equations in integral form, and recombine two more pieces:

$$E = \frac{1}{L}\sum_{i=1}^{2} \int d\beta \left[\log \frac{\rho_i^p + \rho_i^h}{\rho_i^h} \frac{1}{2\pi} \left(\rho_{\bar{i}} - \frac{\mu L}{2\pi} e^{\sigma_i \beta} \right) + \rho_i \log \frac{\rho_i^h}{\rho_i^p + \rho_i^h} \right]$$

We now notice that the first and third pieces contribute in the opposite way when summed over, and so they cancel. In the remaining piece, we can freely sum over i instead of summing over $\bar{i}$, hence we get the desired

$$E = -\frac{\mu}{2\pi} \int d\beta \, e^{\sigma_i \beta} \log(1 + e^{-\varepsilon_i}). \tag{11.3.15}$$

Therefore, in the infrared we get from (11.3.11) that

$$E \sim -\frac{\mu}{2\pi} \sum_{i=1,2} \int_{-\infty}^{\infty} d\beta \, e^{\sigma_i \beta} \log(1 + e^{-\mu L_0 \exp[\sigma_i \beta]}).$$

We can now do the integrals:

$$E \sim -\frac{\mu}{2\pi} \times 2 \times \frac{\pi^2}{12\mu L_0} = -\frac{\pi}{12 L_0}, \tag{11.3.16}$$

which implies an infrared central charge of

$$c_{IR} = -\frac{6 L_0 E}{\pi} \sim \frac{1}{2}, \tag{11.3.17}$$

corroborating the idea that we obtain a free Majorana fermion for the infrared CFT.

The procedure to extract the ultraviolet central charge $c_{UV} = \frac{7}{10}$ is more complicated, and we shall refer to the original paper [2]. It involves the use of the so-called *dilogarithm trick*, which goes beyond the purposes of this simple pedagogical exercise. One needs to study more carefully the TBA equations and find an asymptotic solution for the pseudoenergies when $\mu L_0 < <1$.

References

[1] Fendley P and Saleur H 1992 N = 2 Supersymmetry, Painleve III and exact scaling functions in 2-D polymers *Nucl. Phys.* B **388** 609–26

[2] Zamolodchikov A B 1991 From tricritical Ising to critical Ising by thermodynamic Bethe ansatz *Nucl. Phys.* B **358** 524–46

[3] Fendley P, Saleur H and Zamolodchikov A B 1993 Massless flows. 1. The sine-Gordon and *O(n)* models *Int. J. Mod. Phys.* A **8** 5717–50

[4] Fendley P, Saleur H and Zamolodchikov A B 1993 Massless flows, 2. The exact *S*-matrix approach *Int. J. Mod. Phys.* A **8** 5751–78

[5] Zamolodchikov A B and Zamolodchikov A B 1992 Massless factorized scattering and sigma models with topological terms *Nucl. Phys.* B **379** 602–23

[6] Bazhanov V V, Lukyanov S L and Zamolodchikov A B 1996 Integrable structure of conformal field theory, quantum KdV theory and thermodynamic Bethe ansatz *Commun. Math. Phys.* **177** 381–98

[7] Bazhanov V V, Lukyanov S L and Zamolodchikov A B 1997 Integrable quantum field theories in finite volume: excited state energies *Nucl. Phys.* B **489** 487–531

[8] Bazhanov V V, Lukyanov S L and Zamolodchikov A B 1999 Integrable structure of conformal field theory. 3. The Yang–Baxter relation *Commun. Math. Phys.* **200** 297–324

[9] Bazhanov V V, Lukyanov S L and Zamolodchikov A B 2001 Spectral determinants for Schrodinger equation and Q operators of conformal field theory *J. Stat. Phys.* **102** 567–76

[10] Negro S 2016 Integrable structures in quantum field theory *J. Phys. A* **49** 323006

[11] Ginsparg P H 1988 Applied conformal field theory *Les Houches Summer School in Theoretical Physics: Fields, Strings, Critical Phenomena* (Les Houches: Advanced Study Institute) vol 9

[12] Di Francesco P, Mathieu P and Senechal D 1997 Conformal Field Theory. Graduate Texts *Contemporary Physics* (New York: Springer)

[13] Manolopoulos D and Sfetsos K 2020 *Conformal field theory lecture notes*

[14] Zamolodchikov A B 1990 Thermodynamic Bethe ansatz in relativistic models. Scaling three state Potts and Lee-Yang models *Nucl. Phys.* B **342** 695–720

[15] van Tongeren S J 2016 Introduction to the thermodynamic Bethe ansatz *J. Phys.* A **49** 323005

[16] Fioravanti D, Mariottini A, Quattrini E and Ravanini F 1997 Excited state Destri-De Vega equation for sine-Gordon and restricted sine-Gordon models *Phys. Lett.* B **390** 243–51

[17] Feverati G, Ravanini F and Takacs G 1998 Scaling functions in the odd charge sector of sine-Gordon/massive Thirring theory *Phys. Lett.* B **444** 442–50

[18] Feverati G, Ravanini F and Takacs G 1999 Nonlinear integral equation and finite volume spectrum of sine-Gordon theory *Nucl. Phys.* B **540** 543–86

[19] Ravanini F and Franzini T Thermodynamic Bethe ansatz for a family of scattering theories with $U_q(sl(2))$ symmetry *MSc thesis* (University of Bologna)

IOP Publishing

Integrability using the Sine-Gordon and Thirring Duality
An introductory course
Alessandro Torrielli

Chapter 12

Supplement: a toy model for the Bethe ansatz

In this part of the supplement we display the algebraic Bethe ansatz procedure for a toy model that is closely related to Sine-Gordon at the supersymmetric value of the coupling, though not exactly the same. This model is ideal for pedagogical purposes and to completely highlight the features of the procedure in a simplified setting.

12.1 Setup

The R-matrix that we study here is the one appearing, for instance, in [1] and is relevant for the study of the massless sector of integrable AdS_3 superstring theory [2–33]. We shall employ it here merely as a tool to exemplify the inner workings of the algebraic Bethe ansatz.

The R-matrix is given by

$$R|b\rangle \otimes |b\rangle = |b\rangle \otimes |b\rangle,$$
$$R|b\rangle \otimes |f\rangle = -\tanh\frac{\theta}{2}|b\rangle \otimes |f\rangle + \operatorname{sech}\frac{\theta}{2}|f\rangle \otimes |b\rangle,$$
$$R|f\rangle \otimes |b\rangle = \operatorname{sech}\frac{\theta}{2}|b\rangle \otimes |f\rangle + \tanh\frac{\theta}{2}|f\rangle \otimes |b\rangle,$$
$$R|f\rangle \otimes |f\rangle = -|f\rangle \otimes |f\rangle,$$

(12.1.1)

having set

$$\theta \equiv \theta_1 - \theta_2$$

(12.1.2)

and having denoted by $(|b\rangle, |f\rangle)$ a doublet composed by a boson and a fermion state, respectively. This model therefore has states with statistics given by $(-)^{(0, 1)}$, respectively, as opposed to Sine-Gordon where both states $|1\rangle = |s\rangle$ and $|2\rangle = |\bar{s}\rangle$ (soliton and antisoliton) have the same statistics. The relevant vector space is here therefore $\mathbb{C}^{1|1}$, and the supersymmetry is in this case locally realised [1]. Only in this section of the entire book would one not be able to avoid having to promote the relevant Hopf-algebra formulas to genuine superalgebras.

In one version of the Hopf-superalgebra analysis, the R-matrix is equipped with the following dressing factor:

$$\Phi(\theta) = \prod_{\ell=1}^{\infty} \frac{\Gamma^2\left(\ell - \tau\right)\Gamma\left(\frac{1}{2} + \ell + \tau\right)\Gamma\left(-\frac{1}{2} + \ell + \tau\right)}{\Gamma^2\left(\ell + \tau\right)\Gamma\left(\frac{1}{2} + \ell - \tau\right)\Gamma\left(-\frac{1}{2} + \ell - \tau\right)}, \tag{12.1.3}$$

having set

$$\tau \equiv \frac{\theta}{2\pi i}. \tag{12.1.4}$$

This dressing factor is related to the Sine-Gordon dressing factor at the supersymmetric value of the coupling. Since this value is in the repulsive regime, we can see that the dressing factor does not have any (simple) poles in the physical strip Im $\theta \in (0, i\pi)$, as dictated by the absence of bound states.

Exercise [3 hours' work]: List all the poles and zeroes of $\Phi(\theta)$ in the θ-plane, with their respective orders.

We start by introducing the functions

$$a(\theta) = \text{sech}\frac{\theta}{2}, \qquad b(\theta) = \tanh\frac{\theta}{2}, \tag{12.1.5}$$

in such a way to rewrite (12.1.1) in the following form:

$$R(\theta) = E_{bb} \otimes E_{bb} - E_{ff} \otimes E_{ff} - b(\theta)\left(E_{bb} \otimes E_{ff} - E_{ff} \otimes E_{bb}\right) - a(\theta)\left(E_{bf} \otimes E_{fb} - E_{fb} \otimes E_{bf}\right). \tag{12.1.6}$$

We have the indices of the matrices E_{xy} to denote the states $|b\rangle$ and $|f\rangle$, in a hopefully self-explanatory manner. One thing that is important to remember is that E_{bf} and E_{fb} are fermionic operators because they change the fermionic numbers. We therefore pick up signs when we swap them with other fermionic objects.

We construct the transfer matrix $\text{str}_0 \, T_{(\theta_0|\vec{\theta}\,)}$ over N physical (quantum) spaces and one auxiliary space as the supertrace of the monodromy matrix:

$$T_{(\theta_0|\vec{\theta}\,)} = \prod_{i=1}^{N} R_{0i}(\theta_0 - \theta_i). \tag{12.1.7}$$

The supertrace is appropriate to the superalgebra setting, and is defined as summing over the bosonic states and subtracting over the fermionic states.

As we have previously motivated, we can think of the monodromy matrix as describing the propagation of an auxiliary particle with a rapidity θ_0 past a number of other particles labelled by an index $i = 1, \ldots, N$, each with rapidity θ_i. We shall once again gather all the rapidities of the non-auxiliary spaces into the compact notation $\vec{\theta}$, while the space where the trace is taken is the auxiliary space labelled by the index 0. All of these spaces are isomorphic to $\mathbb{C}^{(1|1)}$ in the end, but it is useful to tag them in different ways for different scopes.

The algebraic Bethe ansatz procedure starts with a choice of the pseudo-vacuum. We know that this is an eigenvector of the transfer matrix that has a special property. The attribute of *pseudo* indicates the fact that this state is not (necessarily) the ground state of the theory—in fact it is only so in rare cases. In our toy model the pseudo-vacuum can be chosen to be the state

$$|0\rangle_p \equiv |\phi\rangle \otimes \cdots \otimes |\phi\rangle. \qquad (12.1.8)$$

Exercise [1.5 hour's work]: Show (using induction) that the pseudo-vacuum is an eigenstate of the transfer matrix, with eigenvalue given by

$$\mathrm{str}_0 T\left(\theta_0 | \vec{\theta}\,\right) |0\rangle_p = \Lambda\left(\theta_0 | \vec{\theta}\,\right) |0\rangle_p, \qquad \Lambda\left(\theta_0 | \vec{\theta}\,\right) = \prod_{i=1}^{N} \Phi\left(\theta_0 - \theta_i\right) \left[1 - \prod_{i=1}^{N} b(\theta_0 - \theta_i)\right]. \quad (12.1.9)$$

It is useful to decompose the monodromy matrix in the auxiliary space using the standard basis:

$$T\left(\theta_0 | \vec{\theta}\,\right) = E_{bb} \otimes A\left(\theta_0 | \vec{\theta}\,\right) + E_{bf} \otimes B\left(\theta_0 | \vec{\theta}\,\right) + E_{fb} \otimes C\left(\theta_0 | \vec{\theta}\,\right) + E_{ff} \otimes D\left(\theta_0 | \vec{\theta}\,\right). \quad (12.1.10)$$

We have split the 0th space as being the leftmost, which leaves us with A, B, C and D being operators on the space $i = 1, \ldots, N$—namely, acting on $[\mathbb{C}^{1|1}]^{\otimes N}$. We can now see very explicitly that the crucial feature of the pseudo-vacuum is being annihilated by C.

Exercise [1.5 hour's work]: Prove that C annihilates the pseudo-vacuum.

This means that we can simulate the operator solution of the harmonic oscillator in ordinary quantum mechanics. We have a state that is an eigenstate of the transfer matrix (which is the generating function of the conserved charges, including the Hamiltonian), and such state is annihilated by the lower diagonal component of a certain matrix of operators—we can now try to create excitations with the upper component.

In fact, we make the familiar ansatz that the eigenvectors of the transfer matrix are obtained by creating M magnons using the B operator:

$$|\beta_1, \ldots, \beta_M\rangle = \prod_{n=1}^{M} B\left(\beta_n | \vec{\theta}\,\right) |0\rangle_p. \qquad (12.1.11)$$

To show that $|\beta_1, \ldots, \beta_M\rangle$ truly is an eigenvector of the transfer matrix for any value of $M \in \mathbb{N}$, one needs to do a certain amount of work. The key is to use the RTT relations which A, B, C and D have to satisfy, as a consequence of R satisfying the Yang–Baxter equation. Such relations play an analogous role as the simple

$[a, a^\dagger] = 1$ relation, which allows us to construct the spectrum of the quantum mechanical harmonic oscillator. We remind that the RTT relations read

$$R_{00'}(\theta_0 - \theta_0') \, T\left(\theta_0|\vec{\theta}\right) T\left(\theta_0'|\vec{\theta}\right) = T\left(\theta_0'|\vec{\theta}\right) T\left(\theta_0|\vec{\theta}\right) R_{00'}\left(\theta_0 - \theta_0'\right). \quad (12.1.12)$$

where we have specified two distinct auxiliary spaces 0 and 0', and the usual N physical (quantum) spaces. By repeated use of the identity

$$a(\theta)^2 + b(\theta)^2 = 1 \quad (12.1.13)$$

we can deduce for instance from (12.1.12) the relation

$$A\left(\theta_0|\vec{\theta}\right) B\left(\theta_0'|\vec{\theta}\right) = \frac{a(\theta_0 - \theta_0')}{b(\theta_0 - \theta_0')} B\left(\theta_0|\vec{\theta}\right) A\left(\theta_0'|\vec{\theta}\right) - \frac{1}{b(\theta_0 - \theta_0')} B\left(\theta_0'|\vec{\theta}\right) A\left(\theta_0|\vec{\theta}\right), \quad (12.1.14)$$

and also an identical one where D replaces A. Recall that we need to show

$$\text{str}_0 T|\beta_1, \ldots, \beta_M\rangle = \text{str}_0 T \prod_{n=1}^{M} B\left(\beta_n|\vec{\theta}\right)|0\rangle_p = \Lambda_M\left(\theta_0|\vec{\beta} \, |\vec{\theta}\right)|\beta_1, \ldots, \beta_M\rangle \quad (12.1.15)$$

for some eigenvalue Λ_M. But (12.1.14) is sufficient to enable us to bring the action of the transfer matrix

$$\text{str}_0 T = A - D \quad (12.1.16)$$

past all the M B's one by one, so that $A - D$ is eventually brought in contact with $|0\rangle_p$ (exactly as we try to bring the harmonic oscillator Hamiltonian through an array of $a^\dagger$ in quantum mechanics). At that point we already know that $|0\rangle_p$ is an eigenstate of the transfer matrix, so we would be done. In the process of swapping $(A - D)$s and Bs we simply accumulate a part that is a genuine eigenvalue, but we also accumulate an unwanted term:

$$\text{str}_0 T\left(\theta_0|\vec{\theta}\right)|\beta_1, \ldots, \beta_M\rangle = \Lambda_M\left(\theta_0|\vec{\beta} \, |\vec{\theta}\right)|\beta_1, \ldots, \beta_M\rangle + X,$$

$$\Lambda\left(\theta_0|\vec{\beta} \, |\vec{\theta}\right) = \prod_{i=1}^{N} \Phi(\theta_0 - \theta_i)\left[1 - \prod_{i=1}^{N} b(\theta_0 - \theta_i)\right] \prod_{n=1}^{M} \frac{1}{b(\beta_n - \theta_0)}. \quad (12.1.17)$$

One has again used various identities between $a(\theta)$ and $b(\theta)$ in order to cancel certain terms in the intermediate steps.

The final part consists in acknowledging that ony when $X = 0$ would we claim that $|\beta_1, \ldots, \beta_M\rangle$ has the property of being an eigenstate of the transfer matrix. It is easy to see that X gathers all the contributions coming from the first term appearing on the right-hand side of the relation (12.1.14). Therefore, with a little thought after writing out this unwanted term explicitly, one sees that one can enforce $X = 0$ by imposing the set of *auxiliary* Bethe equations

$$\prod_{i=1}^{N} b(\beta_m - \theta_i) = 1, \qquad \forall \, n = 1, \ldots, M. \quad (12.1.18)$$

Following the philosophy that we have exhibited earlier, we then impose the quantisation of the transfer-matrix eigenvalue

$$\Lambda(q_1, \dots, q_M; p_k | p_1, \dots, p_{K_0}) = \prod_{i=1}^{K_0} \Phi(\theta_k - \theta_i) \prod_{i=1}^{M} \coth\frac{\beta_i - \theta_k}{2}, \qquad (12.1.19)$$

where we have used the fact that one of the products disappears because of the presence of a multiplying 0, namely $\tanh\frac{\theta_k - \theta_k}{2}$, where one of the particles is taken to be the same as the circulating one. We then end up with the system

$$e^{iL_0 p(\theta_k)} \prod_{i=1}^{K_0} \Phi(\theta_k - \theta_i) \prod_{i=1}^{M} \coth\frac{\beta_i - \theta_k}{2} = 1, \qquad k = 1, \dots, K_0, \qquad (12.1.20)$$

as a set of quantisation conditions for the rapidities of the particles. The momenta are denoted as $p(\theta_k)$, and L_0 is the length of the circle over which the spatial direction is compactified. This applies whenever the system is put in an eigenstate of the transfer matrix (hence of the Hamiltonian, but also simultaneously of all the commuting charges) given by M magnons (flips from $|b\rangle$ to $|f\rangle$) with rapidities β_i, the latter in turn subject to (12.1.18), over a pre-existing layer of N particles all initially polarised as $|b\rangle$.

The solution is given in two steps (or more in systems with a more complicated internal structure). As we have previously pointed out, this procedure goes under the name of *nesting*.

12.2 Low N eigenstates

It is instructive to see explicitly in some simple cases how the procedure that we have just outlined allows us to find the eigenstates of the transfer matrix.

12.2.1 $N = 2$

In this case we can proceed by brute-force diagonalisation to cross-check our results: we need to diagonalise

$$\mathrm{str}_0 T_n = \mathrm{str}_0 R_{01}(\theta_0 - \theta_1) R_{02}(\theta - \theta_2). \qquad (12.2.1)$$

By the index n we intend to signify that we are disregarding the dressing factor from now on because it will be completely straightforward to reinsert it at the end after the diagonalisation procedure has taken place. The dressing factor just comes in front of the R-matrix as a multiplier, and we need to insert one for each of however many particles we consider for every value of N in a simple way. We make this choice for the sake of lightening up the formulas.

The bosonic (overall graded 0) eigenstates can easily be found

$$\mathrm{str}_0 T_n |\phi\rangle \otimes |\phi\rangle = (1 - b_{01}b_{02})|\phi\rangle \otimes |\phi\rangle, \qquad \mathrm{str}_0 T_n |\psi\rangle \otimes |\psi\rangle = (-1 + b_{01}b_{02})|\psi\rangle \otimes |\psi\rangle, \quad (12.2.2)$$

where

$$a_{ij} \equiv a(\theta_i - \theta_j), \qquad b_{ij} \equiv b(\theta_i - \theta_j). \qquad (12.2.3)$$

The fermionic (overall graded 1) eigenstates read:

$$\text{str}_0 T_n\left(|\phi\rangle \otimes |\psi\rangle \pm e^{\pm\frac{\theta_1-\theta_2}{2}}|\psi\rangle \otimes |\phi\rangle\right) =$$

$$\left[\pm e^{\pm\frac{\theta_1-\theta_2}{2}}a_{01}a_{02} + b_{01} - b_{02}\right]\left(|\phi\rangle \otimes |\psi\rangle \pm e^{\pm\frac{\theta_1-\theta_2}{2}}|\psi\rangle \otimes |\phi\rangle\right). \tag{12.2.4}$$

Exercise [2 hour's work]: Write explicitly A, B, C, D in this case and verify the set of eigenstates that has just been given.

We can observe that these eigenstates do not bear any dependence on the auxiliary variable θ_0. This is a consequence of integrability. In fact, the transfer matrix by construction commutes with itself when taken at different values of the auxiliary parameter, as it ought to given that it generates the tower of commuting charges in involution.

For N physical spaces the tower will have N independent commuting charges, one of which will be the Hamiltonian. We have not specified which of the charges is the actual Hamiltonian because we are just using this as a toy model—any of the commuting charges would be a valid Hamiltonian in different physical situations.

The Algebraic Bethe ansatz matches what we find by brute force. It is straightforward to check that the pseudo-vacuum $|\phi\rangle \otimes |\phi\rangle = |0\rangle_p$ is an eigenstate with eigenvalue given by (12.1.17) for $M = 0$. If we then consider the auxiliary Bethe equations, then they are easily solved:

$$b(\beta - \theta_1)b(\beta - \theta_2) = 1, \qquad \text{i.e. } \beta = \pm\infty. \tag{12.2.5}$$

We have to plug in these solutions in the 1-magnon state

$$B(\beta|\theta_1, \theta_2)|0\rangle_p. \tag{12.2.6}$$

We therefore obtain

$$B(\beta|\theta_1, \theta_2)|0\rangle_p = -a(\beta - \theta_2)\left(|\phi\rangle \otimes |\psi\rangle + \frac{a(\beta - \theta_1)b(\beta - \theta_2)}{a(\beta - \theta_2)}|\psi\rangle \otimes |\phi\rangle\right). \tag{12.2.7}$$

which for $\beta = \pm\infty$ gives the result that we know to be the correct one:

$$B(\pm\infty|\theta_1, \theta_2)|0\rangle_p \propto \left(|\phi\rangle \otimes |\psi\rangle \pm e^{\pm\frac{\theta_1-\theta_2}{2}}|\psi\rangle \otimes |\phi\rangle\right). \tag{12.2.8}$$

There is at most a 2-magnon eigenstate, obtained by acting with $B(\infty|\theta_1, \theta_2)B(-\infty|\theta_1, \theta_2)$ onto the pseudo-vacuum. This generates a state that is proportional to $|\psi\rangle \otimes |\psi\rangle$: we have reached the end of the vector space (with $N = 2$ we can only have four eigenstates). If we consider

$$b(\pm\infty) = \pm 1, \tag{12.2.9}$$

then we can check that (12.1.17) is verified—once more by use of (12.2.4) and of standard identities involving the functions a and b.

12.2.2 $N = 3$

In this case the brute-force solution is very cumbersome and we simply use the much leaner way guaranteed by the algebraic Bethe ansatz. The auxiliary Bethe equations read

$$b(\beta - \theta_1)b(\beta - \theta_2)b(\beta - \theta_3) = 1 \qquad (12.2.10)$$

with solutions

$$\beta = \infty, \qquad e^{\frac{\beta}{2}} = -e^{-i\frac{\pi}{2} \pm i\frac{\pi}{4}} \sqrt{\frac{z_1 z_2 z_3}{|\vec{z}|}}, \qquad (12.2.11)$$

having set

$$\vec{z} = (z_1, z_2, z_3), \qquad z_i = e^{\frac{\theta_i}{2}}. \qquad (12.2.12)$$

We indicate with a symbol y the solution having a $+$ sign in the second equation in (12.2.11), in such a way that the solution having a minus sign will be denoted by $-iy$. The β value differ between the two by $i\frac{\pi}{2}$, and the solutions localise on two lines in the complex plane of the rapidity. A solution is colloquially called a *Bethe root*, and sometimes the real part of the Bethe root is called its *centre*.

The formula (12.1.17) can also be recast in an alternative way. This involves a part that is M-independent and another part that depends on M: this latter part just removes a zero and adds another. Specifically we define

$$\mu = e^{\frac{\theta_0}{2}}, \qquad (12.2.13)$$

and write

$$\Lambda(\theta_0|\vec{\theta}) = \frac{-2|\vec{z}|}{\sum\limits_{i=1}^{3}(\mu^2 - z_i^2)} (\mu^2 - y^2)(\mu^2 + y^2) \prod\limits_{n=1}^{M} \frac{1}{b(\beta_n - \theta_0)} \qquad (12.2.14)$$

$$\equiv \Delta(\mu)(\mu^2 - y^2)(\mu^2 + y^2) \prod\limits_{n=1}^{M} \frac{1}{b(\beta_n - \theta_0)}.$$

We see the $N = 3$ zeroes, which, using the variable μ^2, include one zero at infinity. We also see that

$$b(\beta - \theta_0) = 1, \qquad b(\beta - \theta_0) = \frac{y^2 - \mu^2}{y^2 + \mu^2}, \qquad b(\beta - \theta_0) = \frac{y^2 + \mu^2}{y^2 - \mu^2}, \qquad (12.2.15)$$

respectively, in correspondence with the three solutions (12.2.11). The eigenvalues follow this pattern:

$$
\begin{aligned}
&M = 0: && \text{aux. roots none} && \text{eigenv. } \Delta(\mu)(\mu^2 - y^2)(\mu^2 + y^2), \\
&M = 1: && \text{aux. roots } \infty && \text{eigenv. } \Delta(\mu)(\mu^2 - y^2)(\mu^2 + y^2), \\
&M = 1: && \text{aux. roots } y && \text{eigenv. } \Delta(\mu)(\mu^2 + y^2)^2, \\
&M = 1: && \text{aux. roots } -iy && \text{eigenv. } \Delta(\mu)(\mu^2 - y^2)^2, \\
&M = 2: && \text{aux. roots } (\infty, y) && \text{eigenv. } \Delta(\mu)(\mu^2 + y^2)^2, \\
&M = 2: && \text{aux. roots } (\infty, -iy) && \text{eigenv. } \Delta(\mu)(\mu^2 - y^2)^2, \\
&M = 2: && \text{aux. roots } (y, -iy) && \text{eigenv. } \Delta(\mu)(\mu^2 - y^2)(\mu^2 + y^2), \\
&M = 3: && \text{aux. roots } (\infty, y, -iy) && \text{eigenv. } \Delta(\mu)(\mu^2 - y^2)(\mu^2 + y^2).
\end{aligned}
\tag{12.2.16}
$$

We therefore reconstruct all the $2^3 = 8$ eigenvectors. We should mention that no degeneracy is encountered here because we have what we would call *inhomogeneities* (the θ_i s) along the chain of particles, and this naturally removes the degeneracies. Homogeneous spin-chains would not display such a non-degenerate algebraic Bethe ansatz for instance [34]. We should also remark that the eigenvectors are compactly written in terms of creation operators—this is an improvement on brute-force diagonalisation and it provides a more implicit but much more powerful description.

12.2.3 $N > 3$

Even with the simplification offered by the nested algebraic Bethe ansatz, it is nontrivial to write down the solutions in a sufficiently explicit form for arbitrary N. If we focus on $N = 4$ for instance, we continue to find Bethe roots $(y, -iy)$ in addition to the roots located at infinity. Moving on to $N = 5$, a new pair of Bethe roots is found, so that we ought to write $(y_1, -iy_1)$, $(y_2, -iy_2)$, now with two independent centres. With the root at infinity this makes five—it is easy to see that, in order not to overcount, one can restrict the domain to always have exactly N independent Bethe roots.

By forming all possible combinations using *distinct* Bethe roots out of the five, we populate the entire vectors space: $\sum_{M=0}^{5}\binom{5}{M} = 32 = 2^5$.

The eigenvalues are once again organised into the formula

$$
\begin{aligned}
\Lambda(\theta_0|\vec{\theta}) &= \frac{-2|\vec{z}|}{\displaystyle\sum_{i=1}^{5}(\mu^2 + z_i^2)} (\mu^2 - y_1^2)(\mu^2 + y_1^2)(\mu^2 - y_2^2)(\mu^2 + y_2^2) \prod_{n=1}^{M} \frac{1}{b(\beta_n - \theta_0)} \\
&\equiv \Delta'(\mu)(\mu^2 - y_1^2)(\mu^2 + y_1^2)(\mu^2 - y_2^2)(\mu^2 + y_2^2) \prod_{n=1}^{M} \frac{1}{b(\beta_n - \theta_0)}.
\end{aligned}
\tag{12.2.17}
$$

It is still true, as it was for $N = 3$, that the expression involves a polynomial having five zeroes (expressed in the useful variable μ^2 and taking into account the zero at infinity), while different choices of M auxiliary roots work by removing and adding zeroes. For example:

$M = 0$:   aux. roots none   eigenv. $\Delta'(\mu)(\mu^2 - y_1^2)(\mu^2 + y_1^2)(\mu^2 - y_2^2)(\mu^2 + y_2^2)$,

$M = 1$:   aux. roots ∞ eigenv. $\Delta'(\mu)(\mu^2 - y_1^2)(\mu^2 + y_1^2)(\mu^2 - y_2^2)(\mu^2 + y_2^2)$,

$M = 1$:   aux. roots y_1 eigenv. $\Delta'(\mu)(\mu^2 + y_1^2)^2(\mu^2 - y_2^2)(\mu^2 + y_2^2)$,

$M = 1$:   aux. roots $- iy_1$ eigenv. $\Delta'(\mu)(\mu^2 - y_1^2)^2(\mu^2 - y_2^2)(\mu^2 + y_2^2)$,   (12.2.18)

et cetera,

$M = 3$:   aux. roots $(y_1, -iy_1, y_2)$ eigenv. $\Delta'(\mu)(\mu^2 - y_1^2)(\mu^2 + y_1^2)(\mu^2 + y_2^2)^2$,

$M = 3$:   aux. roots $(y_1, -iy_1, -iy_2)$ eigenv. $\Delta'(\mu)(\mu^2 - y_1^2)(\mu^2 + y_1^2)(\mu^2 - y_2^2)^2$,

et cetera.

The form in which the Bethe equations are written respect the algebraic structure, which in this case is based on a rank-one algebra, which is a deformation of $\mathfrak{su}(1|1)$. In general, one can write the Bethe equations almost straightforwardly by staring at the Dynkin diagram of the symmetry group and by relying on the classic treatment of [35, 36].

References

[1] Bombardelli D, Stefański B and Torrielli A 2018 The low-energy limit of AdS₃/CFT₂ and its TBA *J. High Energy Phys.* **10** 177

[2] Babichenko A, Stefanski B Jr and Zarembo K 2010 Integrability and the AdS(3)/CFT(2) correspondence *J. High Energy Phys.* **03** 058

[3] Sfondrini A 2015 Towards integrability for AdS₃/CFT₂ *J. Phys.* A **48** 023001

[4] Borsato R 2015 Integrable strings for AdS/CFT *PhD Thesis* Utrecht U

[5] Sundin P and Wulff L 2012 Classical integrability and quantum aspects of the AdS(3) × S(3) × S(3) × S(1) superstring *J. High Energy Phys.* **10** 109

[6] Ohlsson Sax O and Stefanski B Jr 2011 Integrability, spin-chains and the AdS3/CFT2 correspondence *J. High Energy Phys.* **08** 029

[7] Abbott M C and Aniceto I 2015 Macroscopic (and microscopic) massless modes *Nucl. Phys.* B **894** 75–107

[8] Abbott M C and Aniceto I 2016 Massless Lüscher terms and the limitations of the AdS₃ asymptotic Bethe ansatz *Phys. Rev.* D **93** 106006

[9] Abbott M C and Aniceto I 2021 Integrable field theories with an interacting massless sector *Phys. Rev.* D **103** 086017

[10] Borsato R, Ohlsson Sax O and Sfondrini A 2013 A dynamic $\mathfrak{su}(1|1)^2$ S-matrix for AdS₃/CFT₂ *J. High Energy Phys.* **04** 113

[11] Borsato R, Ohlsson Sax O and Sfondrini A 2013 All-loop Bethe ansatz equations for AdS3/CFT2 *J. High Energy Phys.* **04** 116

[12] Borsato R, Ohlsson Sax O, Sfondrini A, Stefański B Jr and Torrielli A 2013 The all-loop integrable spin-chain for strings on AdS₃ × S³ × T⁴: the massive sector *J. High Energy Phys.* **08** 043

[13] Borsato R, Ohlsson Sax O, Sfondrini A and Stefanski B Jr 2014 The complete AdS₃ × S³ × T⁴ worldsheet S matrix *J. High Energy Phys.* **10** 066

[14] Borsato R, Ohlsson Sax O, Sfondrini A and Stefanski B Jr 2015 The AdS₃ × S³ × S³ × S¹ worldsheet S matrix *J. Phys.* A **48** 415401

[15] Borsato R, Ohlsson Sax O, Sfondrini A, Stefański B Jr and Torrielli A 2017 On the dressing factors, Bethe equations and Yangian symmetry of strings on AdS₃ × S³ × T⁴ *J. Phys.* A **50** 024004

[16] Rughoonauth N, Sundin P and Wulff L 2012 Near BMN dynamics of the AdS(3) x S(3) x S (3) x S(1) superstring *J. High Energy Phys.* **07** 159

[17] Beccaria M, Levkovich-Maslyuk F, Macorini G and Tseytlin A A 2013 Quantum corrections to spinning superstrings in $AdS_3 \times S^3 \times M^4$: determining the dressing phase *J. High Energy Phys.* **04** 006

[18] Sundin P and Wulff L 2013 Worldsheet scattering in AdS(3)/CFT(2) *J. High Energy Phys.* **07** 007

[19] Bianchi L, Forini V and Hoare B 2013 Two-dimensional S-matrices from unitarity cuts *J. High Energy Phys.* **07** 088

[20] Frolov S and Sfondrini A 2022 Massless S matrices for AdS3/CFT2 *J. High Energy Phys.* **04** 067

[21] Frolov S and Sfondrini A 2022 Mirror thermodynamic Bethe ansatz for AdS3/CFT2 *J. High Energy Phys.* **03** 138

[22] Frolov S and Sfondrini A 2022 New dressing factors for AdS3/CFT2 *J. High Energy Phys.* **04** 162

[23] Hoare B and Tseytlin A A 2011 Towards the quantum S-matrix of the Pohlmeyer reduced version of $AdS_5 \times S^5$ superstring theory *Nucl. Phys.* B **851** 161–237

[24] Hoare B 2015 Towards a two-parameter q-deformation of $AdS_3 \times S^3 \times M^4$ superstrings *Nucl. Phys.* B **891** 259–95

[25] Fontanella A and Torrielli A 2019 Geometry of massless scattering in integrable superstring *J. High Energy Phys.* **06** 116

[26] Fontanella A, Ohlsson Sax O, Stefański B Jr and Torrielli A 2019 The effectiveness of relativistic invariance in AdS$_3$ *J. High Energy Phys.* **07** 105

[27] Fontanella A and Torrielli A 2016 Massless sector of AdS$_3$ superstrings: a geometric interpretation *Phys. Rev.* D **94** 066008

[28] Stromwall J and Torrielli A 2016 AdS$_3$/CFT$_2$ and q-Poincaré superalgebras *J. Phys.* A **49** 435402

[29] Pittelli A, Torrielli A and Wolf M 2014 Secret symmetries of type IIB superstring theory on AdS$_3 \times S^3 \times M^4$ *J. Phys.* A **47** 455402

[30] Prinsloo A, Regelskis V and Torrielli A 2015 Integrable open spin-chains in AdS3/CFT2 correspondences *Phys. Rev.* D **92** 106006

[31] Prinsloo A 2014 D1 and D5-brane giant gravitons on AdS$_3 \times S^3 \times S^3 \times S^1$ *J. High Energy Phys.* **12** 094

[32] Regelskis V 2016 Yangian of AdS3/CFT2 and its deformation *J. Geom. Phys.* **106** 213–33

[33] Abbott M C, Murugan J, Penati S, Pittelli A, Sorokin D, Sundin P, Tarrant J, Wolf M and Wulff L 2015 T-duality of Green-Schwarz superstrings on AdS$_d \times$ S$^d \times$ M^{10-2d} *J. High Energy Phys.* **12** 104

[34] Levkovich-Maslyuk F 2016 *The Bethe ansatz. J. Phys.* A **49** 323004

[35] Kirillov A N and Reshetikhin N Y 1986 Bethe ansatz and combinatorics of Young tableaux *Zap. Nauchn. Sem. POMI* **155** 65–115

[36] Kirillov A N and Reshetikhin N Y 1990 Representations of Yangians and multiplicities of occurrence of the irreducible components of the tensor product of representations of simple Lie algebras *J. Sov. Math.* **52** 3156–64

Printed in the USA
CPSIA information can be obtained
at www.ICGtesting.com
CBHW081239060824
12632CB00004B/11

9 780750 358972